Oxford Physics Series

Oxford Physics Series

---

1. F.N.H.ROBINSON: *Electromagnetism*
2. G.LANCASTER: *d.c. and a.c. circuits*
3. D.J.E.INGRAM: *Radiation and quantum physics*
4. J.A.R.GRIFFITH: *Interactions of particles*
5. B.R.JENNINGS and V.J.MORRIS: *Atoms in contact*
6. G.K.T.CONN: *Atoms and their structure*
7. R.L.F.BOYD: *Space physics; the study of plasmas in space*
8. J.L.MARTIN: *Basic quantum mechanics*
9. H.M.ROSENBERG: *The solid state*
10. J.G.TAYLOR: *Special relativity*
11. M.PRUTTON: *Surface physics*
12. G.A.JONES: *The properties of nuclei*
13. E.J.BURGE: *Atomic nuclei and their particles*
14. W.T.WELFORD: *Optics*
15. M.ROWAN-ROBINSON: *Cosmology*
16. D.A.FRASER: *The physics of semiconductor devices*

D.A. FRASER

The Physics of semi-conductor devices

CLARENDON PRESS. OXFORD

1977

Oxford University Press, Walton Street, Oxford OX2 6DP

OXFORD LONDON GLASGOW NEW YORK
TORONTO MELBOURNE WELLINGTON CAPE TOWN
IBADAN NAIROBI DAR ES SALAAM LUSAKA ADDIS ABABA
KUALA LUMPUR SINGAPORE JAKARTA HONG KONG TOKYO
DELHI BOMBAY CALCUTTA MADRAS KARACHI

Casebound ISBN 0 19 851826 9

Paperback ISBN 0 19 851827 7

Printed in Great Britain by Thomson Litho Ltd., East Kilbride, Scotland.

Editor's Foreword

Physics, and many other sciences, would be paralysed without electronic devices. An understanding of what these devices can and cannot do, is essential for the systems engineer and the computer scientist, and the physical principles on which such devices depend need to be known by physicists, electronicists, electrical engineers, and many chemists and materials scientists.

The treatment is basically that of a second-year degree course, but some teachers may prefer to present part of the material in the third year rather than confine it to an integrated course. In keeping with the intentions of the series, this volume is complete in itself as far as the subject matter allows.

The first two chapters of the book define and develop the relevant concepts and relations, before considering individual devices. Wherever possible, the theory is presented in a sufficiently general way to make it of lasting value, not limited by the patented semiconductor products at present in use. The methods are developed with the aid of many diagrams, which often approach the problems from alternative points of view. Manufacturing processes are not treated in undue detail, but the industrial significance of demands for reliability, low noise, and other desirable features is recognized and evaluated.

Overlap with other texts in the Oxford Physics Series is limited and serves to show the need for any applied topics to be based on an adequately wide and thorough understanding of physics. _Atoms in contact_ (OPS 5, Jennings and Morris) and _d.c. and a.c. circuits_ (OPS 2, Lancaster) will give help in the study of semiconductor physics and Rosenberg's _The Solid state (OPS 9) has obvious relevance. Electromagnetism (OPS 1, Robinson) and Radiation and quantum physics (OPS 3, Ingram) provide the sort of introductory 'underpinning' that is necessary for enjoyable learning (and teaching!) at subsequent levels of applied physics.

E.J.B.

Preface

This book concentrates on the use of physical concepts to elucidate the working of semiconductor devices, thereby deepening the reader's understanding of physical ideas. It is therefore expounding an aspect of applied physics. The material that is included has been presented to students in courses ranging from first-year undergraduate level to M.Sc level, but the general standard is that of a second- or third-year undergraduate. A familiarity is assumed with the physical results of a quantum description of matter and with electrical circuit theory (see OPS 2 and OPS 5).

There is no single route through the book. The reader can start with a subject of interest to him, and refer back as required, but it may be noted that the last two chapters contain separate developments of the topics in the first three.

The Problems (at the end of each chapter) vary considerably in standard, being graded from those which require only the insertion of numerical values into the right formula to those which outline a substantial programme of work. In addition, there are requests for action by the reader scattered through the text. These usually require some manipulation of equations, and can help to familiarize the reader with the mathematics, as well as providing a change from straight reading. The author has a personal predeliction for pictorial explanations, and this may be apparent. An ability to sketch the energy-band diagram for each situation should be a target for the reader.

The comments of the author's colleagues on content and presentation have been of great value, and are acknowledged with pleasure.

Chelsea College, London, 1976 D.A.F.

Contents

1. Electrons in solids

Reasonable questions to ask about solids that conduct electricity are, where are the electrons that carry the current? How quickly and in what directions are they going? Some of these questions have straightforward answers, which are given in this chapter: others take much time and paper, and we can only move part of the way to a full answer.

Four descriptions of solids are commonly used in discussing semiconductors. The most sweeping - the equivalent-circuit description - deals only with relations between the terminal voltages and currents in a complete device. Such a description can be regarded as a summary of the understanding gained by the more physical approaches. Of these, the bond model is often too simple to be useful, the energy-band model will be frequently used, and energy/wave-number diagrams will occasionally be needed to explain more subtle processes.

THE BOND MODEL

Most semiconductor devices are made from crystals, and we shall concentrate on these, though noting in passing that non-crystalline glassy semiconductors have recently attracted attention.

Although crystals occur with a marvellous variety both in their external form and the internal patterns of arrangement of their atoms, nearly all important semiconductors have the same crystal structure, that of diamond. From a crystallographer's point of view this is a face-centred cubic structure with eight atoms in the unit cell. A consequence is that a semiconductor is isotropic for many important physical properties, i.e. the property has the same value in all directions. The atoms are arranged so that round any and every atom there are four others making a tetrahedron in space (shown in the case of a Si crystal in Fig.1.1(a)). In a study of semiconductors this tetrahedral arrangement is very important.

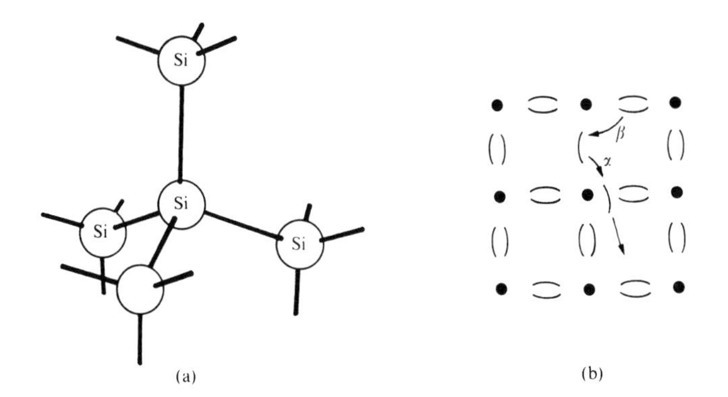

Fig. 1.1. Bonds: (a) each atom in a Si crystal is bound to four others; (b) each bond includes two electrons. At α, an electron has been freed from the bond, allowing it to move, and allowing other transitions, such as β, to occur.

The positions of the electrons relative to the atomic cores can be pinned down by calculation and by X-ray measurements. What one finds is that all except four of the electrons stay close to the nucleus of the atom; the last four, however, tend to be found in the regions directly between two atoms. These four together with the equivalent electrons from neighbouring atoms form the covalent bonds which hold the crystal together, and provide from their number the mobile charges which carry current. The multiple tetrahedral arrangement is not easy to draw, and a flattened pattern is simpler.

In Fig. 1.1(b) one electron is shown displaced from its proper position. Such an electron can move through the crystal and thus carry current. In addition there is a space where it used to be. That space cannot in itself carry current, but it allows other electrons the opportunity of moving without having to be given enough energy to break out of their bond.

As another electron moves into the space, it leaves another space behind it, and so on. It is easier to focus attention on the moving space for an electron rather than a succession of different electrons. The moving space is usually called a hole.

Thus in the bond picture electrical currents can be carried by both electrons and holes simultaneously.

ENERGY BANDS

First we shall explain what is meant by energy bands, then discuss why nature works this way.

Each electron in a solid has a certain total amount of energy, made up of kinetic and potential energy. In a particular sample of a solid there are some ranges of total electron energy where electrons can be found, and other ranges where there are no electrons (Fig.1.2(a)). Close investigation shows that at fairly high values of total energy, there are ranges where, though electrons are rare, they can occur. The complete energy axis can thus be divided into

(a) forbidden bands - no electrons have these energies:
and (b) allowed bands - there may be electrons at these energies.

Usually the lower bands are full of electrons and the upper ones empty. The allowed and forbidden bands come about because

(a) electrons have a wave-like nature;
(b) if the wave equations describing the electrons are to have suitable solutions, some parameters (quantum numbers) must have special values:
(c) (which follows from (b)) an electron has to occupy a 'state', as characterized by a set of quantum numbers;
(d) the regular arrangement of atoms in a crystal leads to electron states having energies only within certain ranges.

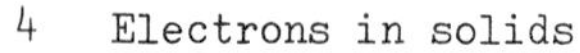

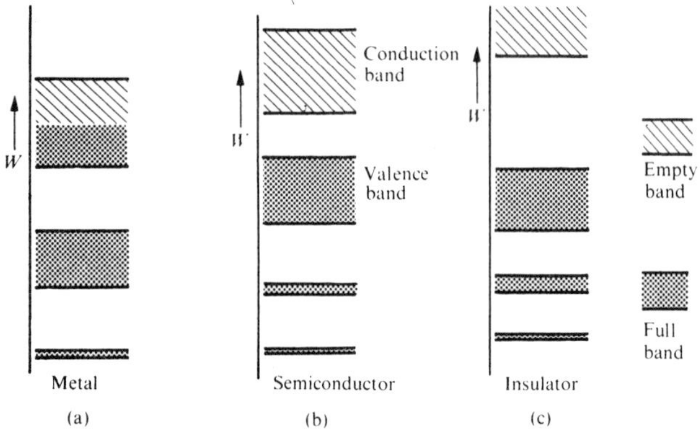

Fig. 1.2. Allowed and forbidden energy bands: (a) in a metal one band is only partly filled; (b) in a semiconductor, the valence band is full (nearly) and the conduction band is empty (nearly); (c) the energy gap between valence and conduction band is larger in an insulator than in a semiconductor.

For many problems, the band picture used without energy values is very helpful. For instance, Fig.1.2 shows how metals, where the highest-energy electrons are in the middle of a band, are different from semiconductors and insulators in which one band, called the valence band, is full, and the next band up, the conduction band, is empty. The magnitude of a forbidden energy band is known as the band gap between the two allowed bands above and below it. The band gap usually referred to is that between the valence and conduction bands: its width is usually measured in electronvolts (eV).

The distinction between insulators and semiconductors is one of degree rather than kind - insulators have larger band gaps, perhaps 3 eV or more, while semiconductors have band gaps from 2·5 eV down to 0·1 eV.

One general point should be noted about energy-band diagrams. The energy axis has plus at the top and minus at the bottom, but because an electron has a negative charge, the electric potential axis, which can also be drawn, has minus at the top and plus at the bottom of the page: be wary.

From a quantum-mechanical point of view each band is finely subdivided, a semiconductor crystal of N atoms having 4N electron states in the conduction band and 4N states in the valence band. N may be as large as 10^{14} for a small transistor, so the subdivisions are very fine. For a more detailed treatment of this point look up the Kronig-Penney model or Mathieu's equation in books on solid-state physics (e.g.Rosenberg, OPS9).

Having considered the possible states that an electron is allowed to occupy, the next question is, which states are occupied? Pauli's exclusion principle says that no two electrons may occupy the same state. At absolute zero temperature the electrons would go into the lowest energy states available. At room temperature, there is a little more energy available for distribution among the electrons, but it is only enough for a small proportion of the electrons in the uppermost levels to be excited a little. For an exact analysis we need to know how many electrons there are at each energy; this is discussed on p.12.

ENERGY/WAVE-NUMBER DIAGRAMS

This level of description takes account of the momentum, and hence the direction of travel, of the electrons as well as their total energy. It is based on a wave picture for the properties of the electrons, and this is derived from wave mechanics (also known as quantum mechanics or matrix mechanics). The wave provides a way of calculating the effects an electron can produce; we can say if we like that the wave is the electron.

Avoiding mathematics, some useful results are:

$$\text{energy} \qquad W = \hbar\omega; \tag{1.1}$$

$$\text{momentum} \qquad p = \hbar k; \tag{1.2}$$

$$\text{group velocity} \quad \upsilon = \frac{d\omega}{dk} = \frac{dW}{dp}, \tag{1.3}$$

where $\hbar$ = Planck's constant $(h)/2\pi$, and ω is the radian frequency of the wave. Eqn (1.1) is familiar from such examples as the photoelectric effect. Eqn (1.2), relating the momentum of a particle (or a wave) to its wave number k (the number of radians of phase change in unit distance), is less often seen, though it is of equal status to eqn (1.1). Eqn (1.3) is the one to use when analysing the motion of a wave packet, which is the way we describe a particle by using waves.

In a solid, the boundaries reflect waves, and if destructive interference is not to occur, the wavelengths must have special values. We look at a simple one-dimensional model, a row of N atoms distance a apart, each with two free electrons, and see what values of k can be used to make allowed wavefunctions, and hence states.

The longest wavelength that can fit our one-dimensional 'crystal' is $2Na$, so the lowest k is given by $\pi/Na = k_1$. The next wave that can fit the boundary conditions has one complete wavelength in the crystal, so that $k_2 = 2\pi/Na = 2k_1$. After that, $k_3 = 3 \times \pi/Na$, and so on, as in Fig 1.3, until $k_{2N} = 2N \times \pi/Na = 2\pi/a = k_a$. We see that all the allowed values of k are equally spaced, and that 2N of them fill the range from 0 to k_a independent of the size of the crystal, and hence the value of N.

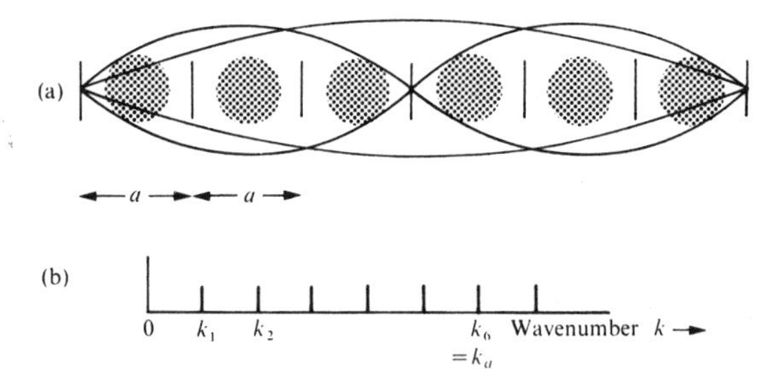

Fig.1.3. Waves in a line of six atoms: (a) the longest two wavelengths are $12a$ and $6a$; (b) the wave numbers of the allowed waves are evenly spaced along the k axis.

The regular spacing of allowed values of k occurs in three directions when a three-dimensional lattice of atoms in ordinary space is analysed. Thus to obtain the view of electron states most appropriate for the start of mathematical analysis, one does well to work in k-space or momentum space. The 2N waves, with k running from k_1 to k_a, are all the waves one needs to make a band. Notice that they are just enough for the number of electrons in the crystal, and that the wavelength goes from the size of the crystal to the distance between atoms.

If one tries to use $k > k_a$, one does not get a new independent solution to the wave equation. This is stated by Floquet's theorem - the Bloch principle. One is allowed to use $k > k_a$ if it is convenient, but the state with wave number k ($> k_a$) is the same as the state which has a wave-number k_a lower.

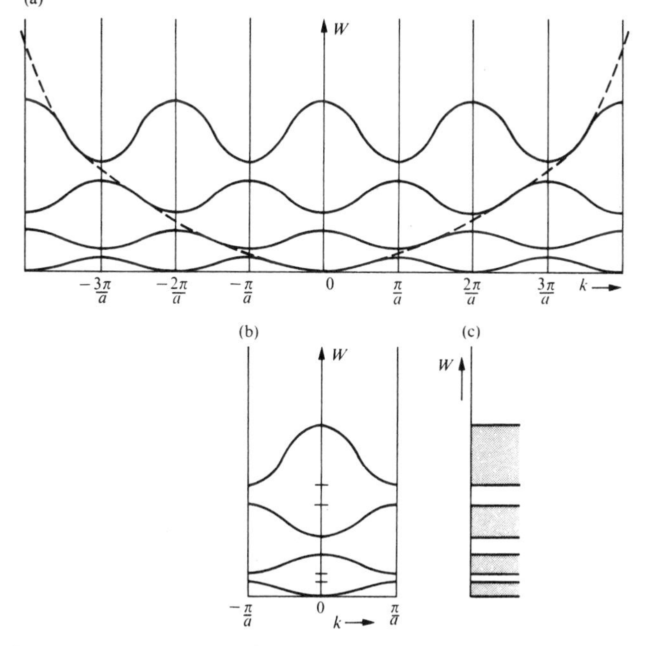

Fig.1.4. Electron energy/wave-number in a periodic potential: (a) W repeats a regular intervals of k; (b) all the information is in the range $\pm\pi/a$; (c) the related energy band diagram.

For another band, other quantum properties of the state have changed, so we may re-use the same wave numbers. The energy of an electron as a function of its wave number is sketched in Fig.1.4(a) for a one-dimensional system. To be precise about electron energy is much more difficult than to be precise about momentum, so Fig.1.4(a) shows only the sort of shape that the curves might have. Remember that the curves should be dotted lines rather than continuous, each dot giving the energy and momentum for a state. One can see that the energy is unaltered if we add k_a to any k - the figure is periodic along the k axis, because the real crystal is periodic in space.

If just one interval of k, of width k_a (a Brillouin zone) is shown as in Fig.1.4(b), the figure is known as the reduced zone diagram. It still contains all the information of Fig.1.4(a), the extended zone diagram.

The allowed and forbidden energy ranges are clear in Fig.1.4(a). In a real semiconductor, we have to specify which direction in the crystal k is along. The W - k curves are different for different directions, and are also of a more complex structure than our simple figure.

EFFECTIVE MASS

Referring back to eqn (1.3), we can write for the acceleration f of an electron due to some external force

$$f = \frac{d\upsilon}{dt} = \frac{d}{dt}\left(\frac{dW}{dp}\right) = \frac{dp}{dt} \cdot \frac{d}{dp}\left(\frac{dW}{dp}\right) .$$

Thus

$$f = \frac{dp}{dt} \cdot \frac{d^2W}{dp^2} .$$

But dp/dt is the rate of change of momentum, and hence equals the applied force P. Thus $(d^2W/dp^2)^{-1}$ replaces the mass in the equation of motion P = mf, and we can describe the response of a carrier to a force by using $(d^2W/dp^2)^{-1}$ instead of the mass.

This new term is known as the *effective mass* (m^*) of a carrier, and summarizes the way interaction with the lattice affects the carrier motion. Explicitly,

$$m^* = \left(\frac{d^2W}{dp^2}\right)^{-1} = \hbar^2\left(\frac{d^2W}{dk^2}\right)^{-1} . \tag{1.4}$$

The effective mass of an electron in the bottom of the conduction band is usually less than the free-electron mass, and may be much less.

The curvature of the top of the valence band can be seen to be of the opposite sign to that of the bottom of the conduction band. The effective mass of an electron at the top of the valence band is therefore negative, though it is more usual to consider holes near the top of the valence band, with effective mass and charge both positive.

DENSITY OF STATES

We can put more detail into our picture of semiconductors by describing the number of electrons at each energy. This will take two steps, first describing the number of states per unit energy range per unit volume - the density of states, and then the fraction at each energy that are filled.

The density of states $S(W)$, as a function of the energy, W is shown in Fig.1.5(a) for an imaginary semiconductor. $S(W)$ is zero in the band gaps, and has a calculable but complex shape in the bands.

A one-dimensional W - k analysis gives shapes for $S(W)$ which are very different from those of a three-dimensional analysis. At this stage our one-dimensional model must be scrapped. Fortunately, only the shapes near the band edges are important, and these are simple. The curves for $S(W)$ spring at right-angles from the W axis, and to a first approximation are parabolas curving either up or down. For the lower edge of the conduction band, we take without proof the result

$$S(W) = (4\pi/h^3)(2m_e)^{\frac{3}{2}}(W - W_c)^{\frac{1}{2}} \tag{1.5}$$

and for the upper edge of the valence band,

$$S(W) = (4\pi/h^3)(2m_h)^{\frac{3}{2}}(W_v - W)^{\frac{1}{2}} \tag{1.6}$$

W_c is the energy of the lower edge of the conduction band, W_v that of the upper edge of the valence band, m_e and m_h are the effective masses of carriers in the two band edges, and h is Planck's constant. Notice that m_e and m_h are the only terms in eqns (1.5) and (1.6) which refer to a given substance.

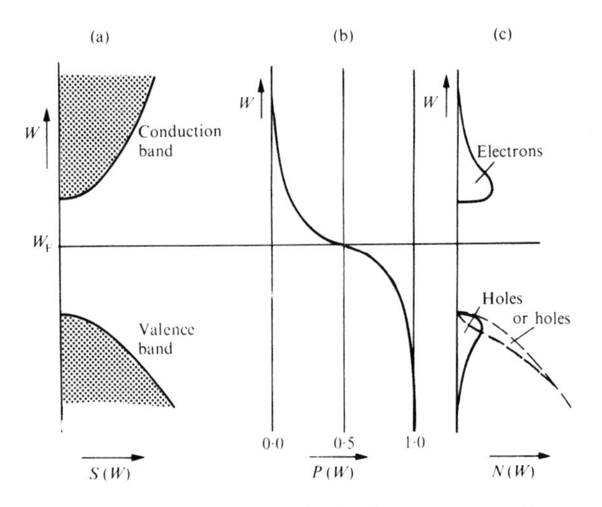

Fig.1.5. To find the number of electrons per unit energy per unit volume in (c), multiply the number of states per unit energy per unit volume in (a) by the probability of the states being occupied in (b). The number of holes is the number of states not occupied, and equals $\{1 - P(W)\}S(W)$.

FERMI-DIRAC STATISTICS

Electrons are found to have the fundamental properties that they are indistinguishable one from another, and that each must be in a distinguishable state. This latter point is known as the Pauli exclusion principle. Particles which have these properties are known as fermions, and are described by Fermi-Dirac statistics. Hence for a system in thermal equilibrium,

$$P(W) = 1/[1+ \exp\{(W - W_F)/\kappa T\}] \qquad (1.7)$$

$P(W)$ is the probability that a state at an energy W is occupied by an electron, κ is the Boltzmann constant, and T is the temperature of the system. We can find where the energy W_F, the Fermi energy or Fermi level must be by noticing that when $W = W_F$, $P(W)$ must equal $\frac{1}{2}$. A suggested definition for W_F is 'The Fermi level is that energy where the chance of a state being occupied by an electron is one half'.

The Fermi level can come in a band gap, but we can do our sums on the chance of a state being filled, even though there are no states at that particular energy. Fig.1.5(b) shows a graph of $P(W)$. Notice the way it is symmetrical about the $\{W = W_F, P(W) = \frac{1}{2}\}$ point, and tends to zero for W very positive, and to one for W very negative. If W is usefully greater than W_F, then there is a simpler approximate form of $P(W)$, since $\exp\{(W - W_F)/\kappa T\} \gg 1$.

$$P(W) \sim 1/[\exp\{(W - W_F)/\kappa T\}] = \exp\{-(W - W_F)/\kappa T\} \qquad (1.8)$$

This form can be derived by assuming that electrons behave like classical micro-billiard balls, and leads to the Maxwell-Boltzmann distribution, but its justification is that for $W \gg W_F$ it is a good approximation to eqn (1.7), derived from non-classical Fermi-Dirac statistics.

To know how many holes there will be in a sample we must know what fraction of states are unoccupied, i.e. we must know $1 - P(W)$

$$1 - P(W) = 1 - 1/[1 + \exp\{(W - W_F)/\kappa T\}]$$

$$1 - P(W) = \exp\{(W - W_F)/\kappa T\}/[1 + \exp\{(W - W_F)/\kappa T\}]$$

When we are interested in holes, the relevant energy range is often well below W_F, and $\exp\{(W - W_F)/\kappa T\} \ll 1$. Then

$$1 - P(W) \sim \exp\{(W - W_F)/\kappa T\} \qquad (1.9)$$

Notice that eqn (1.9) is like eqn (1.8) with a sign reversed. This fits with the idea that holes carry a positive charge, as their energy should then increase as the potential becomes more positive.

ELECTRON AND HOLE DENSITIES

We now have a formula for both S(W) and P(W), and can find the number of electrons N(W) actually occupying states at various energies (Fig.1.5(c)):

$$N(W) = S(W) \times P(W).$$

We can describe the numbers of electrons near the bottom of the conduction band by using the expression for S(W) from eqn (1.5) and the approximate formula for P(W) (eqn (1.8)):

$$N(W) = \frac{4\pi}{h^3}(2m_e)^{\frac{3}{2}}(W - W_c)^{\frac{1}{2}}\exp\{-(W - W_F)/\kappa T\}.$$

The total number of electrons per unit volume over some definite energy range is found by integrating N(W). The number n of electrons in the conduction band is given by

$$n = \int_{W_c}^{W_t} N(W)\,dW,$$

where we integrate from W_c, the energy of the lower edge of the band to W_t at the top of the band. This integral turns out to be awkward, but if W_t is replaced by $+\infty$, the answer is essentially the same, and the answer comes in a standard form:

$$n = \int_{W_c}^{\infty} N(W)\,dW = \frac{2}{h^3}(2\pi m_e \kappa T)^{\frac{3}{2}}\exp\left\{-\frac{(W_c - W_F)}{\kappa T}\right\} \text{ electrons per unit volume} \quad (1.10)$$

The number of holes p in the valence band is found in the same way, using 1 - P(W) instead of P(W) and integrating down from W_v at the top of the valence band:

$$p = \frac{2}{h^3}(2\pi m_h \kappa T)^{\frac{3}{2}}\exp(-(W_F - W_v)/\kappa T) \text{ holes per unit volume} \quad (1.11)$$

If the terms outside the exponentials in (1.10) and (1.11) are collected into single symbols A_c and A_v, where

$$A_c = 2(2\pi m_e \kappa T)^{\frac{3}{2}}/h^3, \quad (1.12)$$

$$A_v = 2(2\pi m_h \kappa T)^{\frac{3}{2}}/h^3,$$

then the expressions for n and p look more compact.

$$n = A_c \exp\{-(W_c - W_F)/\kappa T\}$$
$$p = A_v \exp\{-(W_F - W_v)/\kappa T\} \qquad (1.13)$$

A_c and A_v are called the effective density of states for the conduction band and for the valence band. They can be thought of as the number of states that would be required to give the same value of n (or p) if all the states were at a single energy, that of the band edge. From eqn (1.12) we can see that A_c and A_v vary with temperature, while S(W), the density of states, varies very little.

W_F appears in both the eqns (1.13). It must be the same value both times, so we can solve for W_F or eliminate it.

Multiplying the two parts of (1.13) we get

$$pn = A_c A_v \exp\{-(W_c - W_v)/\kappa T\}.$$

The right-hand side of this equation is dependent only on the temperature and the kind of semiconductor, and not on hole or electron densities, and therefore the left-hand side must be the same value for differently doped (see p.14) samples of the same semiconductor. In intrinsic semiconductors without added impurities, the electron density equals the hole density, and the name ***intrinsic carrier density,*** n_i is given to this carrier density. Thus

$$pn = n_i^2 . \qquad (1.14)$$

This equation is very important in the study of semiconductors, but seems to be difficult to emphasize adequately. For this reason it is suggested that it be called the SEMICONDUCTOR EQUATION. It is an example of the chemical law of mass action. The second result from eqn (1.13) comes from dividing the two equations:

$$p/n = (A_c/A_v) \exp\{(W_c + W_v - 2W_F)/\kappa T\}$$

or

$$W_F = (W_c + W_v)/2 + \kappa T \ln(p/n) + \tfrac{3}{4}\kappa T \ln(m_e/m_h) \qquad (1.15)$$

Eqn (1.15) shows that the Fermi level for intrinsic material (where $\ln(p/n)=0$) is at the average of W_c and W_v, with a small correction when $m_e \neq m_h$. When $p \neq n$, then the Fermi level shifts towards the band with the majority of carriers.

It is worth remembering that the Semiconductor Equation holds for any values of p or n, as long as a thermal equilibrium situation is being described, so it often is valid. However, when p and n are controlled in some device by the external conditions, then we may be far from thermal equilibrium and if so

$$pn \neq n_i^2 .$$

DOPING AND CARRIER DENSITY

Many semiconducting devices are made by introducing small quantities of the proper kind of chemical impurity into the semiconductor lattice - doping the semiconductor.

There are two kinds of dopant, donors and acceptors. A number of pairs of terms that go with donors or acceptors are listed.

TYPE OF DOPANT	DONOR (N_a)	ACCEPTOR (N_d)
Semiconductor type	n-type	p-type
Majority carriers are	electrons	holes
They occur in the	conduction band	valence band
Minority carriers are	holes	electrons
Examples of dopant in Si or Ge	P, As, Sb	B, Al, Ga
Dopant is in group	5	3
Charge on ionized dopant is	+ve	-ve
Fermi level is nearer	conduction band	valence band

The extra electron attached to a donor can be liberated by a small amount of energy, 10-50 meV, so that at room temperature, where $kT \sim 25$ meV it is free to travel over the whole crystal and contribute to an increased conductivity. Acceptors remove an electron from the valence band, leaving a mobile hole.

If both donors and acceptors are added to the same crystal, their effects cancel - only the excess of one above the other has an effect on the free-carrier densities. This is exploited in device manufacture, as an n-type semiconductor can be turned into p-type by adding an excess of acceptors. Indeed the process can be repeated, though three changes between n-type and p-type is about the limit.

If the doping atoms are all ionized, then in an electrically neutral region

$$n + N_a = p + N_d \ , \tag{1.16}$$

where N_a and N_d are the acceptor and donor number density.

Eqn (1.16) in conjunction with $pn = n_i^2$ is sufficient to fix p and n if N_d and N_a are known. If $|N_a - N_d| \gg n_i$ then there are simple solutions. If $N_a > N_d$, the material is p-type, $p = (N_a - N_d)$, and $n = n_i^2/(N_a - N_d)$. If $N_d > N_a$ then the material is n-type, $n = (N_d - N_a)$ and $p = n_i^2/(N_d - N_a)$. It is seldom necessary to deal with situations when $|N_a - N_d| \not\gg n_i$, but if these occur, approximate values for p and n can be obtained by first ignoring the minority-carrier density in (1.16), and then using eqns (1.14) and (1.16) alternately to obtain more accurate solutions. Notice that in doped material, the minority carrier density is much less than n_i.

DOPING METHODS

Doping atoms can be introduced into the semiconductor crystal in several ways. In the growing and purification of the original crystal it is convenient to arrange that its properties will be useful without further doping, at least for part of a device.

A very common method of introducing doping atoms is by diffusion at high temperature from the surface of a sample. We note here that diffusion plays an important part in the physics of semiconductor devices. The doping atoms diffuse to their required positions, charges diffuse while the device is operating, and heat is carried to the heat sink by thermal conduction - another example of diffusion.

Thermal diffusion has been much analysed and one can often take solutions from thermal problems and apply them to the other types of diffusion.

The ease with which a doping atom diffuses into a crystal may be described at different temperatures by a temperature-dependent coefficient of diffusion D:

$$D = D_0 \exp(-\Delta W/kT).$$

The term ΔW is the energy barrier an atom has to surmount on each step of its diffusing journey. It is usually a few electronvolts. Diffusion is carried out at high temperatures - around 1000 K for Si. An hour or so is usually long enough for a stage in a manufacturing process.

In general, a brief diffusion at high temperature will produce a shallow layer, but a long diffusion at a lower temperature is required to diffuse doping atoms deep into a piece of semiconductor. The doping atoms may be supplied to the surface of the sample by a gaseous compound containing the wanted atoms - AsH_3, BH_3, and $AlCl_3$ are examples - or by depositing a solid layer of some substance containing the doping atoms on the surface, and then removing the unwanted excess after the diffusion process has been completed .

Another method of introducing doping atoms is by ion implantation. A beam of ions of energy between 10 keV and 100 ke is fired at the surface. The depth of penetration increases up to $0{\cdot}5 \times 10^{-6}$ m as the beam voltage increases, so that the distribution and the doping density are readily controllable, though elaborate equipment is required in comparison with the diffusion process.

With either diffusion or ion implantation the density of doping atoms tends to be highest at the surface. One way of obtaining a surface layer with few doping atoms and hence of high resistivity is by epitaxial growth. If conditions are just right, atoms arriving at a crystal surface can fit onto the lattice and continue the growth of the crystal (that is what the term epitaxy implies). When the arriving atoms are the same as the bulk semiconductor with no impurities present, then a high-resistivity layer is produced. The growth process is slow, so only thin (10 μm) layers are made in this way.

COLLISIONS

In passing through the crystal lattice, electrons and holes suffer collisions. We study the effect of collisions because they control the diffusion and drift of carriers, and hence the flow of electric current.

The first point to note is that carriers do <u>not</u> collide with the lattice atoms. The allowed wavefunctions have already taken account of the periodic lattice, in a fashion which predicts a plane travelling wave if the periodicity of the lattice is perfect.

Abrupt changes of velocity occur when the lattice is imperfect, and in doped semiconductors two causes of irregularity - phonons and ionized impurities - cannot be avoided.

Phonons are lattice vibrations seen from a quantum point of view. The density of phonons in a solid is found to be proportional to the temperature T. The number of collisions an electron makes in a second is proportional to the distance it travels in a second, and to the density of scatterers, so that the collision frequency for electrons with phonons varies as (thermal velocity)×(phonon density), or as $T^{\frac{1}{2}} \times T$. Consequently the mean time between collision τ_d varies according to

$$\tau_d \propto T^{-\frac{3}{2}} \tag{1.17}$$

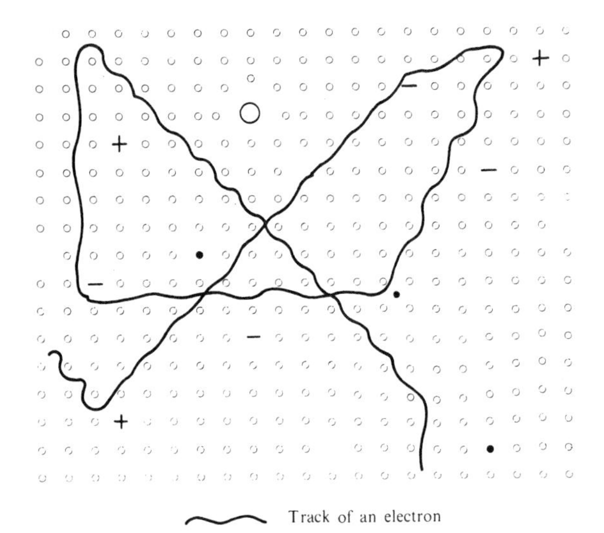

Fig.1.6. This picture is intended to suggest the motion of an electron through a crystal lattice. The wavefunction of the electron fits the regular lattice, so the electron is not scattered from the lattice atoms, but may be scattered from some irregularity. In the picture there are 7 ionized impurities, 2 vacancies, 1 neutral impurity, 1 interstitial atom, and a phonon that can best be seen by viewing the picture from the side.

Scattering from phonons is the most important kind of collision in lightly doped semiconductors at room temperatures.

The frequency of scattering from donor and acceptor ions is proportional to their concentration and is more effective when T is small. Thus ionized-impurity scattering dominates at low T, or in highly doped semiconductors.

Fig. 1.6 shows some of the causes of scattering in a pictorial way, and also attempts to suggest the track of an electron.

CAUSES OF CURRENT

Three reasons for a net flow of holes or electrons are

(a) an electric potential gradient $\frac{dV}{dx}$

(b) a particle-number density gradient $\frac{dn}{dx}$

(c) a temperature gradient $\frac{dT}{dx}$

The last will not be discussed further, though it is important in such useful devices as thermoelectric cooling systems and power generators.

DRIFT CURRENT AND MOBILITY

If an electric field E accelerates carriers of charge q and effective mass m*; the magnitude of their acceleration f is given by

$$f = qE/m^* \ .$$

The distance each carrier is moved in the direction of E by this acceleration is $\frac{1}{2}f\tau^2$, where τ is the time between collisions. The average drift velocity υ_d is found by taking an average value of the distance, and dividing by the average value of τ, thus

$$\upsilon_d = \frac{q}{m^*} \times \frac{\frac{1}{2}\overline{\tau^2}}{\overline{\tau}} \times E$$

We may replace $\frac{1}{2}(\overline{\tau^2}/\overline{\tau})$ by a single symbol τ_d, the mean time between collisions as appropriate to carrier drift, then

$$\upsilon_d = \frac{q\tau_d}{m^*} . E \qquad (1.18)$$

If we define $q\tau_d/m^* = \mu$, known as the ***mobility*** for the particular carriers in the material, then

$$\upsilon_d = \mu E \ .$$

Perhaps the easiest way of thinking of μ is merely as the constant relating υ_d and E. Another way of putting the same idea is that μ is the velocity for unit field.

Notice that, for high μ, we need τ_d high and m^* small.

Consider the drift and current produced by the same field on holes and on electrons. The flux of particles is a useful concept - it is merely the flow of particles without regard for their sign.

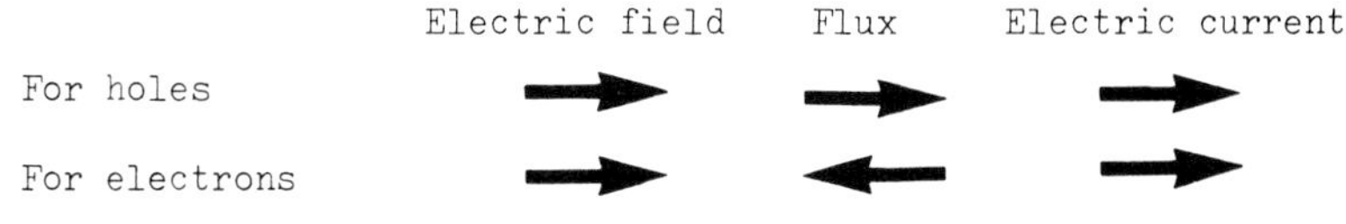

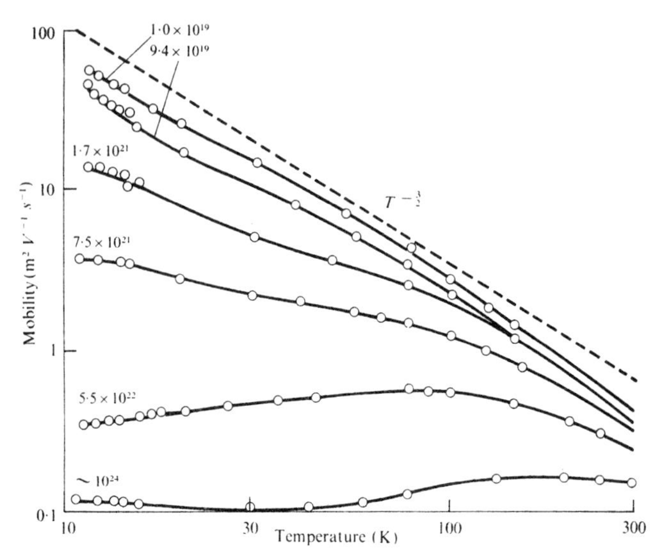

Fig.1.7. Mobility versus temperature for a series of samples of Ge doped with arsenic. The dashed line represents the theoretical slope for scattering by phonons. The number by each curve is the net doping density.(The data is taken with permission from E.M.Conwell(1952).Inst.Radio Engrs.Proc.40,1327-1337).

The drift currents carried by both holes and electrons add to give a larger total if both are present in the same material. The current density j carried by drifting carriers is

$$j = nqv_d,$$

and if there are different kinds of carrier present they contribute separately to the current density.

The conductivity σ of a sample may be defined as the current that flows across unit cube when the field and hence voltage between opposite faces is unity. Thus for a semiconductor where both holes and electrons are present, if we define e to be $+ 1{\cdot}6 \times 10^{-19}$ C,

$$\sigma = ne\mu_e + pe\mu_h \; . \tag{1.19}$$

Mobility is a useful generalizing concept when discussing the effect of electric field on carriers. A very useful first approximation is to say that μ is a constant for a given carrier and material. In fact μ depends on temperature, and on doping density when this is high, because of the effect on the collision time. Fig.1.7 shows these effects. The mobility also depends on the electric field: it falls at high fields, so that the drift velocity in many materials tends to a maximum limit. Nevertheless we shall work with a constant mobility for nearly all devices.

DIFFUSION OF CARRIERS

Diffusion can be looked at in two ways, (a) the independent-particle picture, and (b) the collective-average description.

The independent-particle picture

Here we think of each particle going on its way without regard to the presence or motion of the other diffusing particles; usually this is a valid assumption.

The motion can be described as a series of steps, of a fixed average length ℓ, each of which is in a direction unrelated to the previous step. The style of progress has become known as 'the drunkard's walk'. Einstein showed that, on average, after N steps a particle is a distance $\sqrt{N} \times \ell$ away from its starting position (actually $\sqrt{(3N)} \times \ell$ in a three-dimensional situation). Using the simple $\sqrt{N} \times \ell$ formula, if a particle has to travel 10ℓ then it will take 100 steps. The time taken for a carrier to cross the base of a transistor can be estimated if the step length and thermal velocity are known - now we can see why high-frequency transistors have narrow bases.

The net flow of particles down a density gradient occurs merely because there are more starting from the high-density regions: there is no force exerted by the density gradient forcing particles to diffuse.

The collective average description

This view is not in conflict with the independent particle

picture and predicted results should agree.

The flux F of particles down a density gradient dn/dx is described by

$$F = -D\ dn/dx \qquad (1.20)$$

where D is the diffusion coefficient. Eqn 1.20 is known as Fick's law, and applies to any example of diffusion. The electric current density is

$$J = -qD\ dn/dx,$$

where q is the charge on the carrier.

If we consider the diffusion of holes and electrons in the same density gradient, we see that the hole and electron fluxes are in the same direction, and that the conventional electric currents thus tend to cancel - the opposite of the situation when drift was examined.

	Density gradient	Flux	Electric current
For holes	→	←	←
For electrons	→	←	→

D and μ are related by

$$D/\mu = \kappa T/e\ ,$$

another relation due to Einstein; this relation is derived on p.51. The result is not surprising, as D and μ are each an effect of collisions of carriers with the same lattice defects. Consequently a formula for D is

$$D = \kappa T\ \tau_d/m^*\ . \qquad (1.21)$$

GENERATION AND RECOMBINATION OF HOLES AND ELECTRONS

This is a non-rigorous section, which shows how some ideas can be used to lead quickly to results.

The general principle

The transition rate between two states is the number of transitions per second from one state to the other. In equilibrium, transitions still occur, but the rate one way equals the rate the other way.

For a particular transition the rate is given by a formula

$$K \times (\text{number of candidates}) \times (\text{number of places}) \times \begin{cases} \exp(\Delta W/\kappa T) & \text{for an upward transition} \\ 1 & \text{for a downward transition,} \end{cases} \quad (1.22)$$

where K is a constant. For the upward and downward transitions between two states, the Ks are the same.

In a semiconductor in the conduction band, the number of places = $A_c - n \sim A_c$, and the number of candidates = n. In the valence band, the number of places = p and the number of candidates = $A_v - p \sim A_v$. In this case we think of electrons in the valence band not holes.

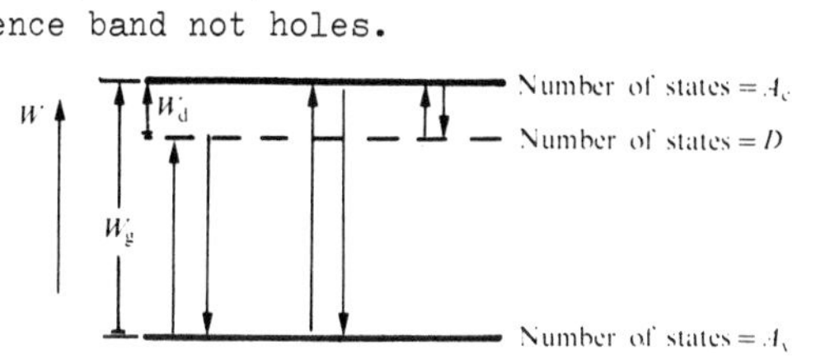

Fig.1.8. Transitions between states in a doped semiconductor. The conduction band and valence band are represented by the equivalent densities of states A_c and A_v. Three pairs of transitions are possible.

Example 1.1. Equilibrium in an intrinsic semiconductor. We may equate the upward and downward transition rates. (The arrows indicate transitions to higher or lower energies).

$$\uparrow A_v A_c \exp\{-(W_c - W_v)/\kappa T\} = \downarrow pn = n_i^2 .$$

Example 1.2. Equilibrium in a doped semiconductor (see Fig.1.8). D is the number of donors and αD is the number of donors not ionized. Consider the equilibrium between donors and the conduction band where W_d is the small energy required to ionize a donor:

$$\uparrow \alpha D A_c \exp(-W_d/\kappa T) = \downarrow n(1 - \alpha)D ;$$

hence

$$\frac{\alpha}{1 - \alpha} = \frac{n}{A_c} \exp(+W_d/\kappa T) ,$$

n/A_c is very small, and $\exp(+W_d/\kappa T)$ is not very large, hence

$\alpha/(1 - \alpha)$ is small and α is small.

Donors are therefore nearly all ionized - as assumed earlier.

Example 1.3. Equilibrium between the conduction band and the valence band.

$$\uparrow A_v A_c \exp(-W_g/\kappa T) = \downarrow np, \tag{1.23}$$

as before. This is the semiconductor equation again.

Example 1.4. Equilibrium between the valence band and donors.

$$\uparrow(1 - \alpha)\, DA_v \exp\{-(W_v - W_d)/\kappa T\} = \downarrow \alpha Dp \tag{1.24}$$

Compare (1.23) and (1.24). Each side of (1.24) has a small term replacing a large one in (1.23), so the transition rates to the valence band are smaller than between the conduction band and the valence band if the Ks are the same.

Example 1.5. Minority carrier recombination. When the carrier concentrations in a semiconductor are disturbed, and the cause of the disturbance is removed, then the concentrations return to normal. We can analyse this process.

Take n-type material to which equal numbers of excess holes and electrons have been added (how?). If equilibrium concentrations are n_0 and p_0 and excess concentrations are n_1 and p_1,

$$\uparrow(\text{transition rate}) = KA_v A_c \exp(-W_g/\kappa T) = K\, n_0 p_0 \,,$$

as before, and

$$\downarrow(\text{transition rate}) = K(n_0 + n_1)(p_0 + p_1)\,.$$

The rate of recombination

$$\downarrow - \uparrow = K\{(n_0 + n_1)(p_0 + p_1) - n_0 p_0\},$$

but in n-type material $n_0 \gg n_1$, so the net rate becomes

$$-dp_1/dt = K\, n_0 p_1 \tag{1.25}$$

i.e. the excess minority carriers decay away exponentially with a lifetime τ_r which is given by Kn_0. In practice τ_r has to be determined by experiment as recombination can occur in several ways, for instance via traps in the middle of the band gap - see pp 82-84.

The average distance that a carrier can diffuse before recombining is known as the ***diffusion distance*** L. If the mean distance between collisions is λ, then in a time τ_r, a carrier can make τ_r/τ_d collisions, and on the independent-particle picture can diffuse a distance $\lambda(\tau_r/\tau_d)^{\frac{1}{2}}$. If the average thermal velocity is $(\kappa T/m^*)^{\frac{1}{2}}$, then eqn (1.21) can be used to show that

$$\lambda(\tau_r/\tau_d)^{\frac{1}{2}} = (D\tau_r)^{\frac{1}{2}} = L\ .$$

THE CONTINUITY EQUATION

The ideas on diffusion, drift and recombination can be combined into a single equation, the continuity equation. This equation makes a useful starting point for further analysis. The equation expresses the idea that the difference between the numbers of particles entering a region and leaving it (no matter how) is equal to the change in the number of particles inside the region.

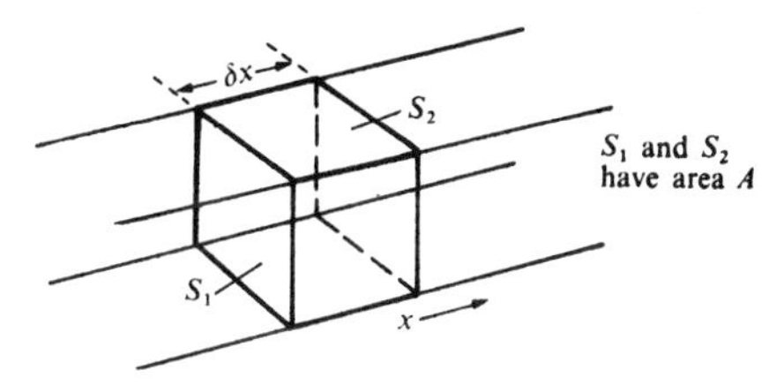

Fig.1.9. Charges enter the element of volume through S_1 and leave through S_2; any difference between the rate of entering and of leaving results in a change in the number contained in the volume.

In Fig.1.9 the number of electrons entering per second by drift and diffusion across the surface S_1, of area A, is

$$-A\mu(nE)_x - AD(dn/dx)_x\ .$$

(We have taken a simple case where the electric field E and the density gradient $\nabla . n$ only have non-zero components in the x direction. The subscript x implies that the quantity is evaluated at x).

The number of particles leaving at the surface S_2 is

$$A\mu(nE)_{x+\delta x} + A\,D(dn/dx)_{x+\delta x}\ .$$

The generation and recombination rates per unit volume are g and r, so the full continuity equation is

$$A\mu(nE)_{x+\delta x} - A\mu(nE)_x + AD(dn/dx)_{x+\delta x} - AD(dn/dx)_x + (g - r)A\,dx = \frac{d}{dt}(nA\delta x)\ .$$

When each term is divided by $A\delta x$, the first four terms may be recognized as differentials as $\delta x \to 0$, so the continuity equation can be written

$$\mu\frac{d}{dx}(nE) + D\frac{d^2n}{dx^2} + g - r = \frac{dn}{dt}\ . \tag{1.26}$$

This is now a differential equation, and boundary conditions and forms for E, g, and r need to be specified before further progress can be made. In later chapters some devices will be analysed - we shall take fairly simple situations, where some terms can be neglected and the rest are simple.

EQUIVALENT CIRCUITS

Of all the general concepts used in electronics, probably only the idea of an equivalent circuit is both applicable to this book and sufficiently difficult to require some discussion. An equivalent circuit provides a representation of some aspect of the electrical behaviour of a device in a form suitable for inclusion in circuit calculations. Different equivalent circuits are needed for describing the way a device responds to small sine-wave signals or to large pulses, for its noise properties or for its d.c. characteristics.

Two common components of equivalent circuits are current and voltage sources (or generators). The circuit function of a source is defined by the symbols beside the diagram. Fig.1.10 shows some examples of such sources: a current generator which supplies a fixed current i_{CB_0}, flowing independently of any

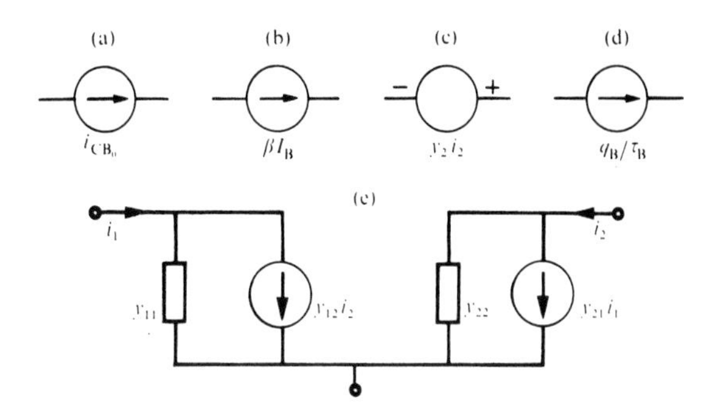

Fig.1.10. Equivalent circuits: (a) a fixed-current generator or current source; (b) a dependent-current generator; (c) a dependent-voltage generator; (d) a current generator dependent on the charge q_B on another circuit element; (e) the y-parameter equivalent circuit for a three-terminal device.

opposing voltage or other circuit parameter (Fig.1.10(a)); a current generator supplying a current βi_B dependent on the value of another current i_B (β is a dimensionless constant)(Fig.1.10(b)); a voltage generator, capable of supplying any current, controlled by a current i_2 via a susceptance y_2 (Fig.1.10(c)); a current generator controlled by a charge q_B (Fig.1.10(d)).

We very often require an equivalent circuit to describe the small-signal properties of devices. Two approaches are possible here - the formal and the physical.

Formal equivalent circuits

We can describe the way a device acts on signals in a circuit by four complex numbers which refer to just one signal frequency. Fig.1.10(e) shows one way of doing this. y_{11}, y_{12}, y_{22} and y_{21} are the four complex numbers; they all have the dimensions of (current/voltage), and as a set are known as the 'y parameters' of the device. If instead of current generators we use voltage generators the result is a set of 'z parameters', while 'h parameters' include both current and voltage generators.

These formal equivalent circuits represent exactly the behaviour of a device at a single frequency. Hence the variation with frequency of each parameter has to be specified (manufacturers on occasion supply graphs of this information). We can infer very little about the physical processes taking place in the device from the formal equivalent circuit.

Physical equivalent circuits

A physical equivalent circuit is based on what is happening in the device, and if skillfully constructed, suggests the physics while, at the same time, representing the behaviour of the device over a wide range of frequency. The description may omit some processes in the interest of simplicity, and hence will not be perfectly accurate. However, extra components can be added to a physical equivalent circuit to model any physical processes deemed worthy of inclusion, and thus (we hope) represent the behaviour more accurately.

A good physical model should therefore be flexible, allowing complexity to be traded for accuracy, and should be helpful in relating features of behaviour to the underlying physical principles.

PROBLEMS

1.1. Draw the graphs of $S(W)$, $P(W)$, and $N(W)$ for p-type semiconductor, showing the Fermi level and the energy level of the acceptors.

1.2. Show that the curve for $N(W)$ has a maximum $\kappa T/e$ above the edge of the conduction band, so long as the doping is not too heavy. Interpret this physically.

1.3. Find the probability that a state is occupied by electrons at a temperature T, when the state is an energy W higher than the Fermi level, for the following cases:

(a)$W = 0{\cdot}01$ eV and $T = 300$ K; (b)$W = 0{\cdot}6$ eV and $T = 300$ K; (c)$W = 0{\cdot}01$ eV and $T = 4$ K. Suggest situations where each calculation might be relevant.

1.4. Calculate the mean free path for electrons and for holes in Si from the mobility. Take the carrier temperature to be 300 K, and express the answer in terms of the distance between Si atoms ($2{\cdot}34 \times 10^{-10}$ m).

1.5. Compounds of the form AB, where A is a group III element and B is a group V element, are semiconductors. By referring to the periodic table, list all such compounds.

1.6. Set up a computer simulation for carrier diffusion. Divide the distance through which carriers move into equal parts (ten is satisfactory). The diffusion process is modelled by arranging that the carriers in region N at time T move in equal numbers to region (N+1) and to region (N-1) at time (T+1). At the boundaries of the whole distance carriers are injected or absorbed as appropriate for the physical situation being modelled. The diffusion of carriers through the base of a transistor is an instructive topic, and thought is required to relate computed times and distances to physical times and distances.

2. Bulk effects

Although most semiconductor devices exploit the properties of interfaces, there are a number of uses to which a uniform block of semiconductor can be put, and these form the subject matter of this chapter. The physics of a material is also conveniently studied in samples whose properties are uniform throughout. Chapters 3 and 4 build on the concepts introduced in this chapter, examining the physics and applications of interfaces.

RESISTANCE

The conductivity of a block of semiconductor is (eqn (1.19)

$$\sigma = ne\mu_e + pe\mu_h \ .$$

In semiconductors the conductivity can range between wide limits. The conductivity is a minimum and the resistivity a maximum when the semiconductor is intrinsic, that is when $n = p = n_i$. (Exercise. This is strictly true only when $\mu_e = \mu_h$. Work out the condition for minimum conductivity when $\mu_e \neq \mu_h$.)

For Ge, the maximum resistivity at 300 K is about 0·49 Ω m. The equivalent figure for Si would be several thousand ohm metres, but it has not been possible to obtain pure enough Si, so that in practice several hundred ohm metres is the maximum resistivity.

The maximum conductivity is set by there being a limit to the solubility of doping atoms in the semiconductor lattice, though scattering from ionized doping atoms reduces mobility and the benefit of high doping. Resistivity in Ge and Si can be reduced below $10^{-5}\Omega$ m.

In integrated circuit processing, thin layers are used where the doping varies with depth below the surface. Consequently it is more convenient to work with the surface resistivity S, rather than the varying volume resistivity. S is quoted in ohms per square, as a square of any size has the same resistance between two opposite sides. Values of S in practice might range between 5 Ω per square and 500 Ω per square and it is found convenient to make resistors between 10 Ω and 10^5 Ω. Outside this range, other circuit techniques are used.

The resistance of a block of semiconductor varies with temperature. The effect is put to good use in some devices, and is a ***bad thing*** in others. First we shall examine the dependence of n and p on temperature, then combine this information with the temperature variation of the mobility to give the dependence of σ (or ρ) on the temperature T. Fig.2.1 sketches the variation of majority and minority carrier density for a doped semiconductor. The intrinsic-carrier density increases steadily as T increases. The majority-carrier density is roughly constant over a fair range, and is about equal to the density of the uncompensated doping atoms. At high temperatures, the intrinsic density surpasses the doping density, and both majority and minority densities increase rapidly. At very low temperatures, the doping atoms are no longer ionized, and the majority-carrier density is low. If the majority-carrier density is required to be constant in a device, then the range of temperature marked 'extrinsic' in Fig 2.1 is most satisfactory. Notice that the minority-carrier density depends strongly on temperature over all temperature ranges.

We can examine the rate at which n_i and the minority-carrier density vary with temperature by finding an expression for n_i. Multiply eqn.(1.10) and eqn.(1.11), and take the case when $p = n = n_i$:

$$n_i = \frac{2}{h^3}(2\pi kT)^{\frac{3}{2}}(m_e/m_h)^{\frac{3}{4}}\exp\{(W_c - W_v)/2kT\} \qquad (2.1)$$

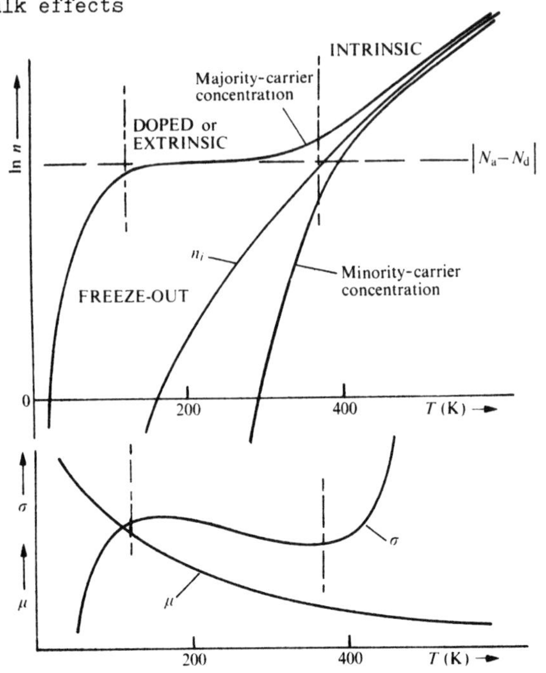

Fig.2.1. The variation with temperature of majority and minority-carrier concentration, and of mobility μ and conductivity σ for a semiconductor.

Differentiating by parts with respect to T, and them simplifying gives

$$\frac{dn_i}{dT} = \frac{n_i}{T}\left(\frac{3}{2} + \frac{W_c - W_v}{2\kappa T}\right)$$

where $W_c - W_v$ is the band-gap energy W_g, so we may write

$$\frac{dn_i}{n_i} = \frac{dT}{T}\left(\frac{3}{2} + \frac{W_g}{2\kappa T}\right).$$

For Si, W_g is 1·1 eV and κT is about 25 meV at 300 K, so that $W_g/2\kappa T \sim 22$, and $dn_i/n_i = (dT/T) \times 23\frac{1}{2}$. At 300 K, if dT is 1 K, then dn_i/n_i is $23\frac{1}{2}/300$ or about 8 per cent, so that n_i in Si at 300 K increases by 8 per cent for a 1 K temperature rise. The minority-carrier density will depend on $(A - D)/n_i^2$, so it will vary by 16 per cent for 1 K temperature change. In Ge

W_g is smaller, and the minority-carrier density only varies by 10 per cent for dT = 1 K.

If we consider semiconductors that are not too heavily doped, at temperatures around 300 K (this covers a lot of cases), then mobility (μ) varies roughly as $T^{-\frac{3}{2}}$. The variation of μ is only detectable as a change in σ when the majority-carrier density is not changing. Over the 'extrinsic' range of temperature (see Fig.2.1), the resistance of a semiconductor increases with temperature, because μ is falling. Nevertheless, the common statement that the resistance of semiconductors falls as temperature rises is not wrong, if we interpret it as applying only to intrinsic semiconductors.

Thermistors are semiconductor resistors whose temperature dependence is exploited; they are available with either a positive or negative temperature coefficient of resistance. A thermistor can be used as a transducer to provide an electrical signal describing the temperature of its surroundings, or as a non-linear circuit element, where the resistance changes as the electrical power dissipated in the thermistor is varied. In either use, the thermistor responds only to slow changes, perhaps below 1 Hz.

THE HALL EFFECT

When a current of density $\underline{J}$ flows in a semiconductor perpendicular to a magnetic field $\underline{B}$ (Fig.2.2), there is a force on the holes and electrons perpendicular to both $\underline{B}$ and $\underline{J}$, and a transverse current tends to flow. The process is known as the ***Hall effect,*** and provides a way of estimating the majority-carrier density and sign, and is also the basis of devices for measuring magnetic field and for multiplying two signals.

In Fig.2.2, if J_x and B_z are as shown, then because holes and electrons have oppositely directed drift velocities and opposite signs of charge, both of them will tend to be accelerated in

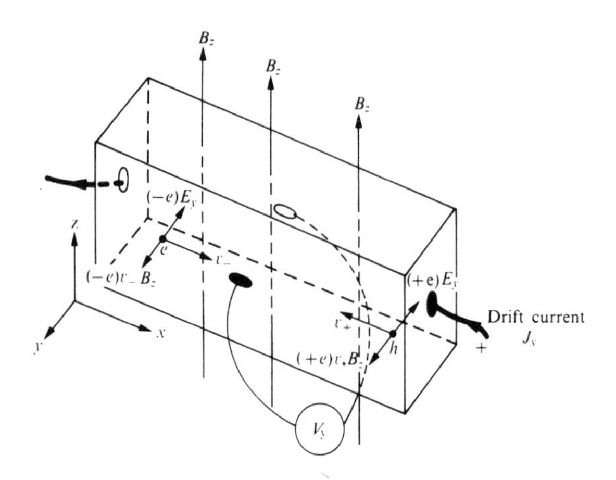

Fig.2.2. The Hall effect in a semiconductor. Electrons and holes drift in opposite directions, but the transverse $(\pm e)v_{\pm}B_z$ force is in the same direction for both of them.

the direction of the positive y axis by a force qv_xB_z, where v_x is the drift velocity of the carriers, and is given by

$$J_x = N_{ma}qv_x \,.$$

N_{ma} and q are the concentration and charge of the majority carriers (we are dealing with cases where the minority carriers are too few to be evident).

If the circuit is arranged so that the transverse current J_y is zero, the transverse force qv_xB_z must be balanced by a transverse electric force qE_y so we can write

$$qE_y = qv_xB_z = J_xB_z/N_{ma} \,. \tag{2.2}$$

Eqn (2.2) is often rearranged,

$$1/qN_{ma} = J_xB_z/E_y = R,$$

where R is called the Hall constant for the sample. It is negative for n-type semiconductors and for simple metals, and positive for p-type semiconductors, where the majority carriers are holes.

The three quantities J_x, B_z, and E_y may all be measured so that N_{ma} can be determined. A measurement of resistivity on the same sample gives μN_{ma}, and thus in combination with the Hall measurement yields a value for the mobility of the majority carriers. The use of the Hall effect to measure magnetic field is straightforward. A sample has a fixed current passed through it, and is calibrated in a known magnetic field. InSb is a useful material for a magnetic field probe, as it has a high electron mobility.

If the magnetic field is produced by current flowing in a coil, eqn (2.2) shows that the output E_y is proportional to the product of the value of B_z, produced by the coil current, and J_x. Thus the output is a constant times the product of two currents - a result there are few other ways of achieving.

The transverse force on an electron ($-ev \times B$) tends to make it move in a circle. If it makes no collisions on the way, the angular frequency of rotation is given by

$$eB/m = \omega_c$$

the electron cyclotron frequency. (Check this using $F = mv^2/r$ for motion in a circle).

If τ_d is the electron collision frequency, the product $\omega_c \tau_d$ may be written, using the definition of μ following eqn(1.18);

$$\omega_c \tau_d = \frac{eB}{m^*} \tau_d = \frac{e\tau_d}{m^*} B = \mu B \; .$$

For a magnetic field to have a large effect on the electrical properties of a semiconductor, the circular motion of the electrons must be interrupted only seldom by collisions. Hence a simple criterion for a high magnetic field, in a semiconductor context, is

$$\text{either} \quad \omega_c \tau_d \gg 1 \qquad \text{or} \quad \mu B \gg 1 \; .$$

If a magnetic field is parallel to the flow of current it can still have an effect on carrier motion. The effect is known as magnetoresistance - the resistance increases when the magnetic field is present.

PHOTOCONDUCTIVITY

The basic process of photoconductivity is the absorption of photons by the semiconductor, resulting in the production of free carriers. Photoconductors exploit the resulting change of conductance. When the photon raises an electron from the valence band to the conduction band (Fig 2.3(a)), producing both a free electron and a hole, we have intrinsic photoconduction. When the photon ionizes a donor or acceptor, we have impurity photoconduction (Fig 2.3(b)). In the latter case, only one free carrier is produced, and the energy required is much less than that of the band gap of the semiconductor.

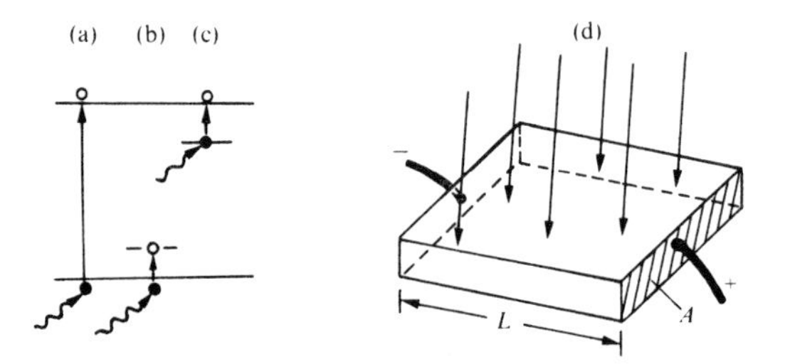

Fig.2.3. Photoconductivity: (a) intrinsic; (b) and (c) impurity photoconduction; (d) light falls on the photoconductor and produces C electron-hole pairs.

The impurity semiconductors are used in an unusual state. They are cooled to a low enough temperature for the impurities not to be ionized, until such time as energy is provided by the arrival of photons. The ionization energy of Au acceptors in Ge is 0·15 eV, equivalent to 9 μm wavelength. Au-doped photoconductors are cooled with liquid nitrogen to ensure that the Au is not ionized. Zn-doped Ge (ionization energy 0·03 eV) is used to detect 40 μm far infrared radiation, and is cooled to liquid helium temperatures. Thus the general rule is that the longer the wavelength of the radiation, the lower the temperature to which the detector has to be cooled.

In passing through an absorbing medium a beam of light is attenuated according to the law

$$dn = n_0 \exp(-x/\lambda) \times (1/\lambda) \times dx = (1/\lambda) n(x) dx.$$

The chance of a photon being absorbed in a small distance dx is thus dx/λ. In intrinsic photoconductors, every atom can absorb, and λ may be 10^{-8} - 10^{-6} m (how many atomic diameters is this?). In impurity photoconductors, only the impurity atoms can absorb; there may be few of these, and λ may be much longer. A way of increasing the chance of a photon being absorbed is to surround the photoconductor with a reflecting cavity, so that photons pass and re-pass through the detector.

The carriers liberated by photons are used to carry current, so we are interested in how long they are available. In Fig 2.3, a beam of light is producing C electron-hole pairs per second in a semiconductor. C will depend on the number of photons arriving at the surface, the fraction of these not reflected, and the fraction of those which are absorbed which produce electron-hole pairs rather than heat.

If the electron and hole lifetimes are τ_e and τ_h in the steady state, the extra carrier numbers (not densities) are

$$\Delta N = C\tau_e \quad \text{and} \quad \Delta P = C\tau_h \ .$$

The change in the conductance Δg equals $\Delta\sigma A/L$ where $\Delta\sigma$ is the change in the conductivity ($\Delta\sigma = e(\Delta N\tau_e + \Delta P\tau_h)/AL$), A is the cross-sectional area of the semiconductor, and L is its length. Thus

$$\Delta g = \frac{e}{L^2}(\Delta N\mu_e + \Delta P\mu_h) = \frac{eC}{L^2}(\mu_e\tau_e + \mu_h\tau_h) \qquad (2.3)$$

For Δg to be large, $\mu_e\tau_e$ and/or $\mu_h\tau_h$ should be large and L should be small. The values of τ_e and τ_h can be adjusted over wide ranges by controlling trap densities, but when τ_e and τ_h are long the frequency response of the device may be poor. To see this consider C to be made up of a steady and a sinusoidal part,

$$C = C_0 + C_1 \exp(j\omega t).$$

We ignore C_0 and the corresponding steady parts of ΔN and ΔP in this a.c. analysis. The rate of change of electron density is due to the varying part of C and to recombination, thus

$$\frac{d\Delta N}{dt} = \frac{-\Delta N}{\tau_e} + C_1 \exp(j\omega t).$$

If we try a solution $\Delta N = \Delta N_1 \exp(j\omega t)$, then

$$\Delta N_1 = C_1 \tau_e/(1 + j\omega\tau_e),$$

with a similar equation for holes. Therefore

$$\Delta g(\omega) = \frac{eC_1}{L^2}\left(\frac{\mu_e \tau_e}{1+j\omega\tau_e} + \frac{\mu_h \tau_h}{1+j\omega\tau_h}\right) \tag{2.4}$$

When $\omega\tau_{e,h} >> 1$, Δg falls as $1/\omega$, so that $1/\tau_{e,h}$ represents a cut-off frequency for the device.

An implicit assumption is that carriers recombine before they drift to the end-contacts of the device. However, even if the carriers do reach the contacts, the equations remain unchanged. Consider two cases:

(a) One kind of carrier is trapped. Then as electrons drift out of the positive end of the semiconductor, more must enter at the negative end to keep the whole device neutral. Only when recombination at a trap occurs can ΔN or ΔP fall.

(b) Neither kind of carrier is trapped. Electrons will still enter at the negative contact and holes at the positive contact to replace the carriers that have drifted away. The excess carrier density can fall only by the recombination process. As a result, each photon may allow many carriers to pass through the photoconductor, especially if it is short. The process is known as photoelectric gain, and, as shown by eqn (2.4) is a demonstration of the way sensitivity may be traded for frequency response.

THE HAYNES-SCHOCKLEY EXPERIMENT

This experiment used the photoproduction of electron-hole pairs to demonstrate the importance of minority carriers, and provided a way of measuring their mobility, complementing Hall measurements which give the mobility of majority carriers.

Electron-hole pairs are produced by a pulsed light source focused onto a small area of semiconductor (Fig 2.4). These carriers drift in opposite directions in an applied electric field. The minority carriers are detected by a reverse-biased metal point contact (see p 115 for an explanation of why a metal point can detect minority carriers). The arrival of the minority carriers is observed on an oscilloscope to be delayed by a time which is proportional to the distance the carriers have travelled, so the minority carriers have a definite drift velocity.

The drift velocity is found to be proportional to the drift field, so the minority-carrier mobility can be calculated. The value of the mobility for a specific carrier (e.g.electrons in Si) turns out to be the same whether they are majority or minority carriers, and the experiment shows that minority carriers are needed for a complete explanation of semiconductor phenomena.

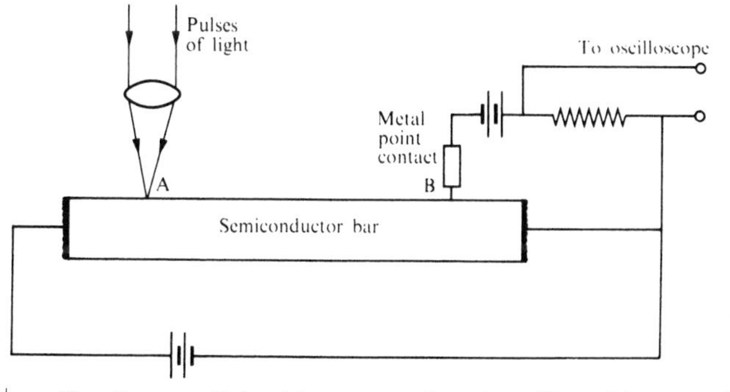

Fig.2.4 The Haynes-Schockley experiment. Minority carriers are produced by pulses of light at A and drift to B where they can be detected.

NEUTRALITY

Because semiconductors contain mobile electric charges, they tend to be electrically neutral, which is to say they contain equal amounts of positive and negative charge. It is interesting to see how large a region can be non-neutral, without there being large potential differences.

Let us consider p-type material with a region where a varying potential tends to a constant value, taken as zero. In the constant potential, the hole density p_0 will equal the acceptor density A. Where the potential has changed to V, the hole density will be controlled by the Maxwell-Boltzmann relation, as in eqn (1.9).

$$p = p_0 \exp(-eV/\kappa T).$$

Poisson's equation for this situation is

$$\frac{d^2V}{dx^2} = \frac{-e}{\epsilon_r\epsilon_0}(p - A) = \frac{-ep_0}{\epsilon_r\epsilon_0}\{\exp(-eV/\kappa T)-1\}$$

where ϵ_0 is the permittivity of free space, and ϵ_r is the relative permittivity of the semiconductor.

This is unpleasant to solve in the general case, but when $|eV/\kappa T| \ll 1$, we can use the first two terms in a series approximation for the exponential, giving

$$\frac{d^2V}{dx^2} = \frac{e^2 p_0}{\epsilon_r\epsilon_0\kappa T} V \, . \tag{2.5}$$

Eqn (2.5) has a solution

$$V = V_0 \exp(-x/\lambda_D), \tag{2.6}$$

where $\lambda_D = (\kappa T\epsilon_r\epsilon_0/e^2p_0)^{\frac{1}{2}}$ and is known as the Debye length. Eqn (2.6) shows that a perturbation in the potential tends to build up or die away over distances of the order of λ_D. For a doping density of $10^{22}\,m^{-3}$ in Si at 300 K, $\lambda_D = 5 \times 10^{-8}$ m.

Major field changes occur over distances greater than λ_D; the most important example of this happening is in the depletion layer in a p-n junction, which is examined on p.51.

SEMICONDUCTOR PLASMAS

Sometimes the electrical properties of a semiconductor cannot be explained by analysing the behaviour of a typical carrier and then multiplying by the number of carriers present. On occasion, it is necessary to consider all the carriers from the start of the analysis; the semiconductor is then said to be behaving as a plasma, and to be exhibiting collective properties. One such property, the tendency to neutrality, has been discussed in the previous section.

Another plasma effect is a collective oscillation at the plasma frequency ω_p, which equals $(ne^2/\epsilon_r\epsilon_0 m^*)^{\frac{1}{2}}$, where n is the concentration of mobile carriers. Consider a layer of charge at a position x, which has been displaced by a distance ξ (Fig 2.5). We assume that there is an equal concentration of immobile charge of the opposite sign to ensure average neutrality. If ξ is not the same for all layers there will be compression or rarefaction, measured by $d\xi/dx$, and leading to non-zero space charge. Thus

$$\epsilon_r\epsilon_0\frac{dE}{dx} = -ne\frac{d\xi}{dx}\ .$$

The acceleration of the layer of charge will be

$$\frac{d^2\xi}{dt^2} = \frac{Ee}{m^*}\ .$$

Differentiating with respect to x,

$$\frac{d}{dx}\left(\frac{d^2\xi}{dt^2}\right) = \frac{d}{dx}\left(\frac{Ee}{m^*}\right) = \frac{-ne}{\epsilon_r\epsilon_0 m^*}\frac{d\xi}{dx}\ .$$

We can change the order of differentiation,

$$\frac{d^2}{dt^2}\left(\frac{d\xi}{dx}\right) = \frac{-ne^2}{\epsilon_r\epsilon_0 m^*}\frac{d\xi}{dx}\ .$$

This is the equation of simple harmonic motion with a frequency $1/(2\pi)(ne^2/\epsilon_r\epsilon_0 m^*)^{\frac{1}{2}}$, and represents the mobile charges

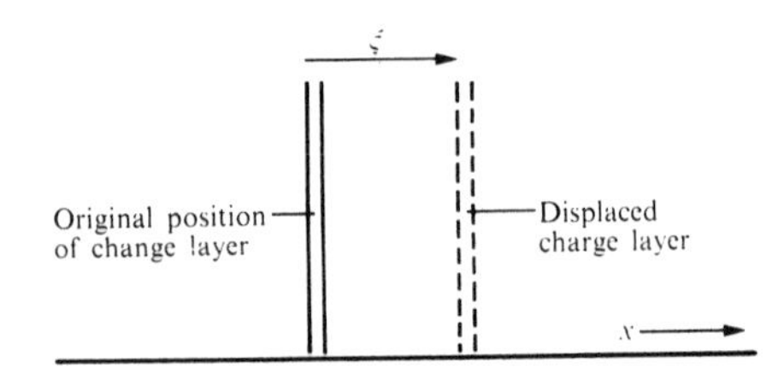

Fig.2.5. Plasma oscillations: the layers of charge are not all equally displaced, so that rarefaction or compression occurs.

performing a longitudinal oscillation.

In this analysis collisions have been ignored, and in semiconductors the collision frequency is often greater than ω_p. As a result, the effect of oscillations at the plasma frequency is seldom directly observable, but is more likely to be involved indirectly in a detailed analysis of the performance of a device. Nevertheless, plasma effects can be directly observed in semiconductors; an example is the propagation of circularly polarized electromagnetic waves (helicon waves) in doped semiconductors in magnetic fields.

GUNN DIODES

A uniformly doped sample of n-GaAs (or some other 3-5 semiconductor) when provided with two end-contacts, can function as a microwave oscillator, and is then known as a Gunn diode. The functioning of a Gunn diode depends on the unusual drift velocity versus field curve shown in Fig.2.6(a), which is a direct consequence of the details of the band structure of GaAs (Fig.2.6(b)). There are minima in the conduction band at two energies, and electrons have very different effective masses and hence mobilities near the two minima. In the lower valley electrons are 'light' and have a high mobility. This is the region normally occupied. When the electrons acquire energy from a high electric field, they tend to be scattered into the higher valley, where their effective masses are much greater

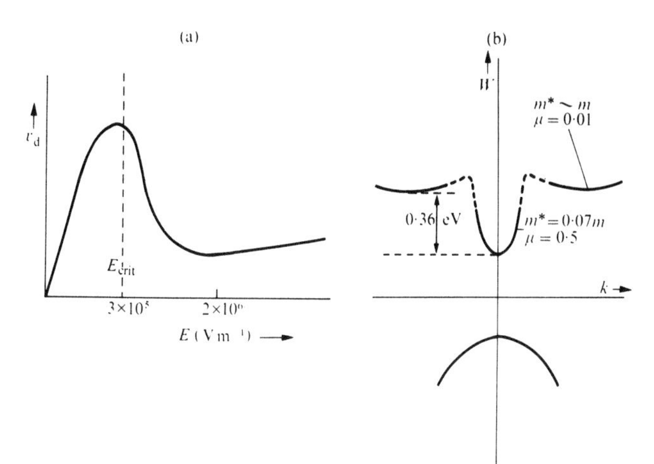

Fig.2.6 GaAs has properties which are exploited in Gunn diodes: (a) the drift velocity (v_d) falls sharply when the electric field (E) rises above E_{crit}: (b) electrons occupy the sharp lowest minimum of the conduction band when $E < E_{crit}$, but are scattered into the higher energy minima when $E > E_{crit}$.

and their mobilities much reduced.

The current through a sample is proportional to the drift velocity, and the voltage across it to field, so Fig.2.6(a) can be regarded as an I-V characteristic for a sample. Notice that there is a region where the differential resistance is negative; this is a strong indication that unstable behaviour and oscillation is likely.

Instability manifests itself in two ways. The applied voltage is not distributed uniformly over the length of the device, and this non-uniform field changes as time goes by. Gunn diodes oscillate in a variety of ways, but we shall consider one mode only for further examination. In this mode a local high-field region forms near the negative electrode (the cathode), and propagates through the diode to the anode (Fig.2.7). The double change of field is associated with a dipole-charge layer, a depletion region in front and an electron-rich region behind

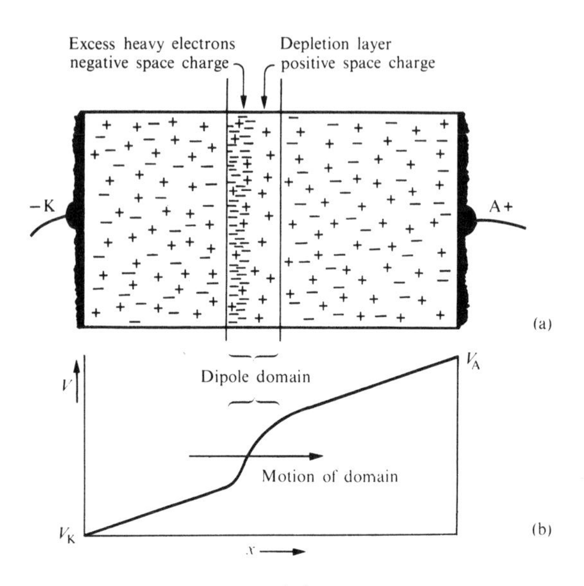

Fig.2.7 Gunn diode domain: (a) A sketch of the dipole domain suggests the depletion layer at the head of the domain followed by the accumulation layer (the positive charges are fixed donor ions). (b) The voltage rises more rapidly in the domain than in the rest of the diode.

(check that this fits with Poisson's equation). The electron-rich layer is in the high field, so the electrons are 'heavy' and move no faster than the 'light' electrons in the low field. The dipole layer and local high field are known as a domain. The domain, having been formed at the negative contact, travels to the positive contact. On reaching it there is an extra pulse of current, the average field rises as there is no high field domain in the diode, and a new domain forms at the cathode to repeat the cycle.

The domain travels with the velocity of the drifting electrons, so that the period of oscillation is ℓ/v_d if the time required for domain formation is neglected (ℓ is the length of the diode, and v_d is the electron drift velocity). The drift velocity is about 10^5 m s^{-1}, so that a Gunn diode 10 μm long should oscillate at 10 GHz.

If this mode of oscillation is to occur, the domain must be usefully smaller than the device. At the leading edge of the domain, the maximum space-charge density will occur if there are no electrons present, and will equal eN_d. If we integrate Poisson's equation across the depletion layer we get

$$\delta \ell e N_d = \epsilon_r \epsilon_0 (E_{max} - E_{min})$$

where $\delta \ell$ is the width of the depletion layer, E_{max} is the high field behind the depletion layer, and E_{min} is the low field in the bulk in front of the depletion layer. E_{max} is about 2×10^5 V m^{-1}, while E_{min} is much less and the exact value is not important. The product $\delta \ell N_d = 10^{14}$ m^{-2}, and thus sets a lower limit to the length of material which can set up a travelling domain and oscillate in this mode. To give the domain a chance to build up and propagate, $\ell > 10\delta\ell$, so we have as a design criterion

$$\ell N_d > 10^{15} \text{ m}^{-2} .$$

The voltage supplied to Gunn diodes has to form both the low-field and the high-field regions. Either may be the larger contribution, so a simple calculation of the voltage will not be attempted. Often a voltage supply between 5 V and 20 V is needed. The current taken by the diode is proportional to its area and can be chosen by the designer. With good heat sinks continuous wave (c.w.) operation is possible, though pulsed operation is more usual. An example has

Power output	Frequency
250 mW c.w.	18 GHz

PROBLEMS

2.1. Calculate the resistance at 300 K of a block of Si of length 0·5 cm and cross-section $3 \times 10^{-7} m^2$ doped with $4 \times 10^{23} m^{-3}$ P atoms. What current flows when 1 V is applied across its length, and what power density is dissipated in its volume?

2.2. Plot a graph of the conductivity at 300 K of Si as the concentration of As is gradually increased from 0 to $2 \cdot 0 \times 10^{22} m^{-3}$ in a sample originally doped with $1 \cdot 0 \times 10^{22} m^{-3}$ atoms of B.

2.3. A photoconductor has an effective width of 10 cm and a length of 1 mm. Light of wavelength $6 \cdot 274 \times 10^{-7} m$ falls onto it at a power density of $1 \cdot 0$ W m^{-2}, generating electron-hole pairs of 100 ns lifetime. The hole mobility is $0 \cdot 001\ m^2 V^{-1} s^{-1}$ and the electron mobility is $1 \cdot 0\ m^2 V^{-1} s^{-1}$. If each photon produces an electron-hole pair, what conductance is produced by the light?

2.4. Find the carrier density for which ω_p for n-Ge equals the collision frequency as calculated from mobility and effective mass. Also, find the magnetic field for which the electron cyclotron frequency equals the collision frequency.

2.5. Find what relation there is between the formulae for the plasma frequency, the Debye length, and the root-mean-square thermal velocity of the carriers in a plasma.

3. p-n junctions

CONSTRUCTION AND DIRECT-CURRENT CHARACTERISTICS

The junction between p- and n-type semiconductors has properties which make it the basis of many electronic devices. Carriers must be able to pass from one side of the junction to the other without suffering recombination, so the semiconductor must not contain more than a small number of imperfections. In practice this means either that the device has been made from a slice cut from a large single crystal, parts of which have been transformed by diffusing doping atoms from the surface, or that new material has been grown epitaxially to extend a crystal substrate and to allow the including of a p-n junction. A large p-n junction might cover 10^{-5} m^2 (a square of side 3 mm), while in integrated circuits a device might cover no more than 10^{-10} m^2 area.

A p-n junction diode can rectify alternating currents, i.e. it can pass currents readily in one direction and not in the other. In Fig.3.1, V_f, the voltage in the forward direction for large currents to start to flow is about 0·2 V for Ge and about 0·6 V for Si. I_f depends on the area and the provision for cooling the diode, and may be 10^{-4} - 10^{+2} A. I_r, the reverse leakage current, can be as much as a few microamperes for Ge diodes at room temperature, but is 1000 times less for Si diodes. The reverse breakdown voltage V_r is under the designers control, and may be as low as a few volts or as high as a kilovolt.

THE p-n JUNCTION IN EQUILIBRIUM

It is of value to analyse a simple situation - a p-n junction in equilibrium with no applied voltage or net current.

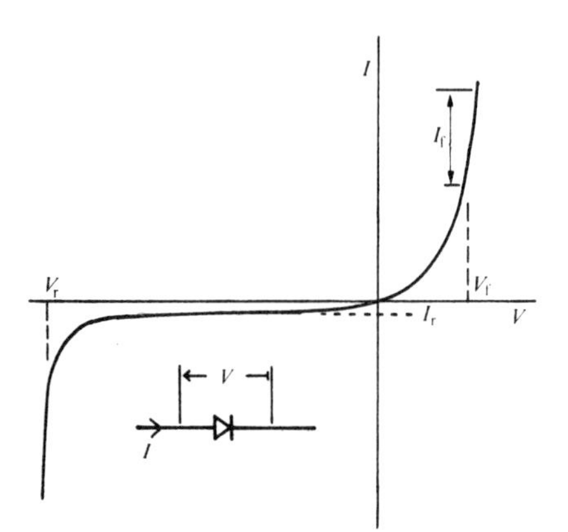

Fig.3.1. The voltage-current characteristics for a p-n junction diode.

Without examining the detailed variation of density or potential across the junction, three statements can be made:

(a) the net current of electrons across the junction is zero,

(b) the net current of holes across the junction is zero,

(c) at all points $pn = n_i^2$.

The zero total current density of electrons can be thought of as two cancelling components caused by drift and diffusion as in (Fig.3.2). The two components may be written as

$$-n(x)e\mu_e \frac{dV(x)}{dx} + eD_e \frac{dn(x)}{dx} = 0 , \tag{3.1}$$

where $V(x)$ is the potential and $n(x)$ the electron density at a distance x from one side of the junction (the cancellation of two currents in equilibrium is used in several places as a starting point for theory).

From eqn (3.1)

$$\mu_e \frac{dV(x)}{dx} = D_e \frac{1}{n} \frac{dn(x)}{dx} . \tag{3.2}$$

Integrate this from $x = x_n$, well on the n-side of the junction, to $x = x_p$ safely on the p-side.

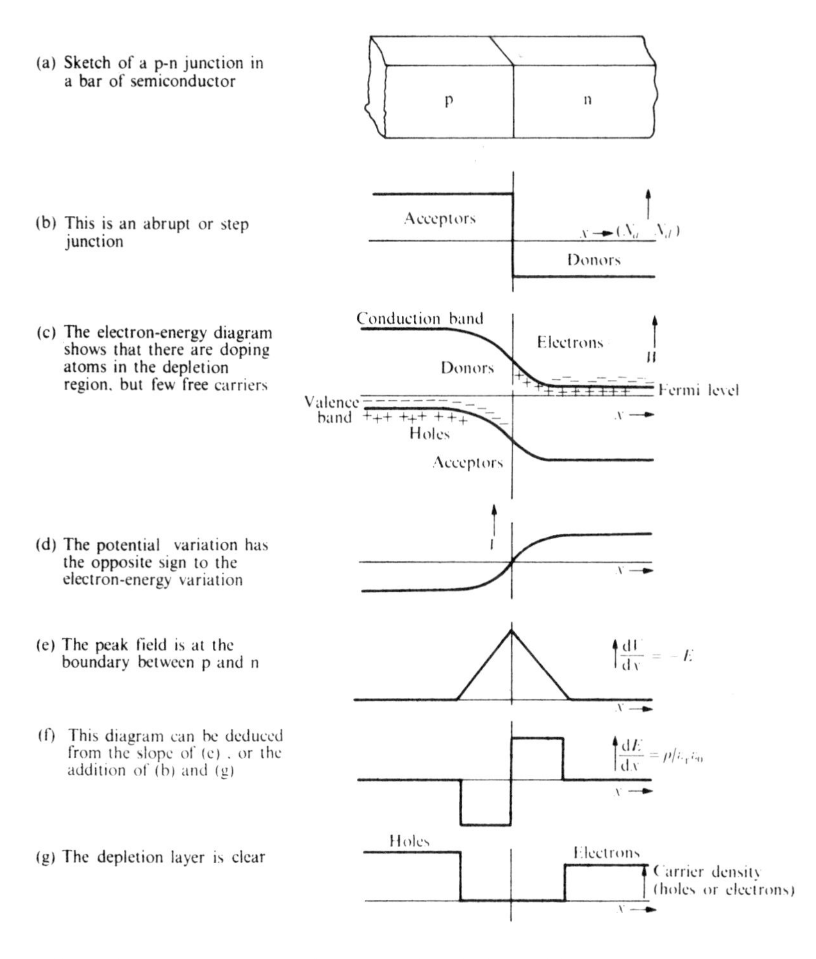

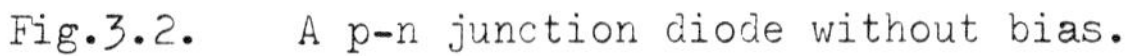
Fig.3.2. A p-n junction diode without bias.

$$\int_{x_p}^{x_n} \mu_e \frac{dV(x)}{dx} dx = \int_{x_p}^{x_n} D_e \frac{1}{n} \frac{dn(x)}{dx} dx$$

The left-hand side becomes an integral in V with limits V_n and V_p, and the right-hand side an integral in n with limits n_n and n_p, in each case well away from the junction.
Then

$$V_p - V_n = (D_e/\mu_e) \ln (n_n/n_p).$$

If the junction was made by doping the n-side with N_d donors and the p-side with N_a acceptors, then to a good approximation

$$n_n = N_d \quad \text{and} \quad n_p = n_i^2/N_a .$$

Hence

$$V_n - V_p = (D_e/\mu_e) \ln (N_d N_a/n_i^2) . \qquad (3.3)$$

There is thus a potential difference between the two sides (referred to from now on as the ***barrier potential*** V_b) which curbs the tendency of the electrons to diffuse away from the place where they are densest. Unless the doping is very heavy, the barrier potential is less than the band gap of the semiconductor. Typical values of barrier potential for junctions in Ge and Si might be 0·4 V and 0·8 V.

In an equilibrium situation the Fermi level is constant right through the system. In Fig.3.2(c) one can see that the difference of the potential energy on the two sides is equal to

$$\{(W_c - W_f) \text{ on p-side}\} - \{(W_c - W_f) \text{ on n-side}\}.$$

Eqn (1.13) gives the form for this,

$$V_b = V_p - V_n = (\kappa T/e)\ln(A_c/n_n) - (\kappa T/e)\ln(A_c/n_p),$$

but, as before $n_n = N_d$ and $n_p = n_i^2/N_a$, so that

$$V_b = (\kappa T/e)\ln(N_d N_a/n_i^2) . \qquad (3.4)$$

If this is compared with eqn (3.3), we can see that

$$\kappa T/e = D_e/\mu_e \ . \tag{3.5}$$

Eqn (3.5) is known as the ***Einstein relation***. The whole derivation can be repeated for holes; the barrier potential comes out the same, and there is a corresponding version of eqn (3.5)

$$\kappa T/e = D_h/\mu_h \ .$$

Thus if either mobility or the diffusion constant is known, the other quantity can be calculated.

DEPLETION LAYERS IN p-n JUNCTIONS

A region close to the metallurgical junction tends to be depleted of holes and electrons, and is known as a ***depletion layer*** (Fig.3.2(g)).

Fig.3.3 shows p-n junctions with forward and reverse bias. Notice that the externally applied bias shows up as a difference between the Fermi levels on the two sides - this is a fundamental point in reading or constructing diagrams. The reverse bias adds to the barrier potential and results in a wider depletion layer. The forward bias subtracts from the barrier potential - in general the forward bias never reverses the usual situation to make the p-side more positive (check with the forward bias voltages and barrier potentials quoted earlier in the chapter.)

So far we have not investigated the detailed structure of the depletion layer. This we now undertake, in a simple case, which is however a reasonable representation of some real devices. We make two assumptions. The first is that the density of doping atoms changes sharply from one value N_a on the p-side to another steady value N_d on the n-side (fig 3.2(b)). Such a junction is known as a ***step junction*** or an ***abrupt junction***. In practice, as long as the change-over occurs in a distance that is much less than the full width of the depletion layer, our

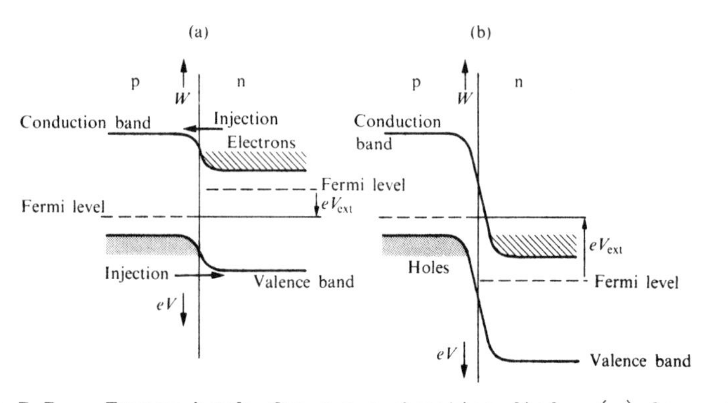

Fig.3.3 Energy bands for a p-n junction diode: (a) forward bias: (b) reverse bias.

assumption will fit.

The second assumption is that the distance over which the electron or hole density falls to a small value is short enough to be ignored. This is equivalent to saying that the depletion layer must be much thicker than the Debye length, or that the total potential difference V_j across the junction is much greater than $\kappa T/e$.

In the depleted parts of the p-region, the only charges are fixed acceptor ions, which are negative. Thus

$$\frac{dE}{dx} = \frac{\rho}{\epsilon_r \epsilon_o} = \frac{-eN_a}{\epsilon_r \epsilon_o} ,$$

where ϵ_o is the permittivity of free space and ϵ_r is the relative permittivity of the semiconductor. Hence

$$E = -(eN_a x/\epsilon_r \epsilon_o) + C .$$

When $E = 0$, $x = x_p$ at the boundary of the depletion layer in the p-region, so that

$$E = eN_a(x_p - x)/\epsilon_r \epsilon_o .$$

Because of our choice of the direction of the x axis in fig 3.2, x_p is a negative number, so E reaches its most negative value E_{max} when $x = 0$,

$$E_{max} = eN_a x_p/\epsilon_r\epsilon_o \ . \tag{3.6}$$

In the depleted part of the n-region, the only charges are the donor ions, which are positive,

$$\frac{dE}{dx} = \frac{eN_d}{\epsilon_r\epsilon_o} \ ,$$

$$E = eN_d(x - x_n)/\epsilon_r\epsilon_o \ ,$$

where x_n is the boundary of the depletion layer in the n-region. From the analysis of the n-layer,

$$E_{max} = -eN_d x_n/\epsilon_r\epsilon_o \ . \tag{3.7}$$

The total potential difference V_j across the junction is the sum of the voltages across the two parts,

$$V_j = \frac{e}{2\epsilon_r\epsilon_o}\,(N_d x_n^2 + N_a x_p^2) \tag{3.8}$$

The two equations for E_{max} (3.6) and (3.7) must give the same answer, so $N_a|x_p| = N_d|x_n|$. Hence there will be a thicker depletion layer where the doping is lighter, thinner where there is heavy doping. Remembering this we can look again at eqn (3.8) and see that most of the voltage appears across the lightly doped side. In an extreme case we can ignore the thickness and voltage on the heavily doped side. Such a junction may be indicated by a plus sign (p^+- n), and the total thickness of the depletion layer is proportional to $V_j^{\frac{1}{2}}$.

Another situation which is straightforward to analyse is a linear graded junction, where the net doping varies linearly near the junction (Fig.3.4). The density of doping atoms is then proportional to the distance x from the metallurgical

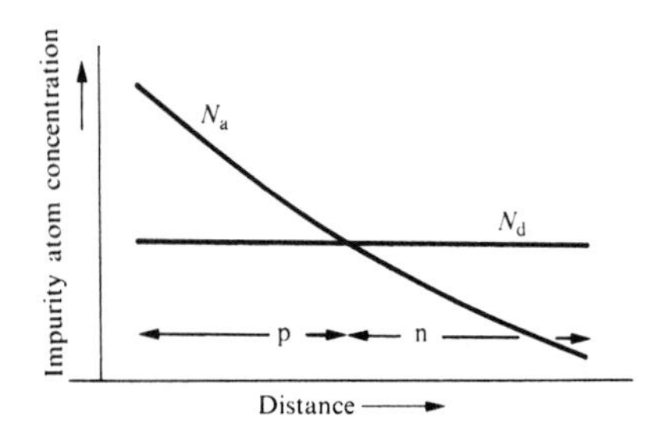

Fig.3.4. The combination of a uniform donor concentration and a variable acceptor concentration can produce a linear graded junction.

junction, E is proportional to x^2, and V to x^3. This kind of junction is often a good approximation to the junctions in planar silicon integrated circuits where doping atoms have diffused in from a surface. Instead of quoting N_a and N_d the junction is described by the gradient of the net doping density.

DEPLETION-LAYER CAPACITANCE

A p-n junction can act as a capacitor. The useful property is the small-signal capacitance $C(V_j)$, which varies with the bias voltage,

$$C(V_j) = \frac{d\, Q(V_j)}{dV_j} = \frac{dQ(V_j)}{dX}\,\frac{dX}{dV_j}\,, \qquad (3.9)$$

where Q is the charge on each side of the capacitor, X is the total thickness of the depletion layer, and V_j is the total difference in potential between the p- and n-sides. If we take a p-n^+ diode, then the major part of V_j and of X is in the p-region. Then $X \sim x_p$ and $Q = N_a e x_p A$ (A is the area of the junction).

Hence

$$\frac{dQ}{dX} = \frac{dQ}{dx_p} = eN_aA\ .$$

From eqn (3.8), when $x_n \ll x_p$,

$$\frac{dX}{dV_j} = \left(\frac{dV_j}{dX}\right)^{-1} = \left(\frac{2e\,N_a x_p}{2\epsilon_r\epsilon_o}\right)^{-1} . \tag{3.10}$$

We can eliminate x_p from (3.10) by using (3.8) again and then write out the expressions for the two terms in eqn (3.9), giving

$$C(V_j) = A(eN_d\epsilon_r\epsilon_o/2)^{\frac{1}{2}}V_j^{-\frac{1}{2}} .$$

Thus $C(V_j)$ decreases as the bias becomes more negative. If $C(V_j)$ is expressed in terms of the depletion-layer thickness (check this using eqn (3.8) once more), then

$$C = A\epsilon_r\epsilon_o/X, \tag{3.11}$$

so the capacity is identical with that of an ordinary capacitor of the same size, shape, and permittivity as the depletion layer.

In our analysis of transistors we shall see how the depletion-layer capacitance is a factor which limits performance. In most diodes the capacitance is unwanted, but in varactor diodes or varicaps, the capacitance is exploited.

Some circuits (including parametric amplifiers) require a capacitor whose value can be controlled by an electrical signal. A reverse-biased p-n junction will do this job admirably - when the capacity is to vary, the bias can be altered. The maximum frequency f at which a varactor diode can be used is set by the time constant $\tau = C(V_j)R$, where R is the resistance of the semiconductor on either side of the p-n junction,

$$f \ll \frac{1}{2\pi\tau} .$$

If a linear graded junction is used instead of a step junction, the relation between capacity and voltage takes the form

$$C(V_j) \propto V_j^{-\frac{1}{3}} \tag{3.12}$$

and other forms (which may be convenient for use in particular circuits) can be obtained by suitably arranging the variation of doping density.

MINORITY-CARRIER INJECTION

When a p-n junction is forward-biased (p made more positive), there is a lowering of the barrier which was preventing the majority carriers from moving across the depletion layer. Consequently there will be a flow of majority carriers across the junction, becoming minority carriers and contributing to a measurable current. We shall first examine conditions at the depletion-layer boundary, and then the way the carriers move away from the junction.

Depletion layers are thin, and carriers can move frequently backwards and forwards across them. As a result there is a local equilibrium between the carrier densities on the two sides of a depletion layer for each kind of carrier separately. If we use a subscript $_p$ or $_n$ to indicate the location of carriers and a subscript $_0$ to indicate equilibrium (e.g. p_{n0} means the density of holes in the n-side in equilibrium), then the relation between densities on the two sides of the junction is

$$p_n/p_p = \exp(-eV_j/\kappa T) = n_p/n_n , \qquad (3.13)$$

which is another application of the Boltzmann relation.

Because neutrality is maintained outside the depletion layers, the majority and minority-carrier densities increase (or decrease) by equal amounts. As a result, while the minority-carrier density can readily change by a large factor, the majority-carrier density changes only by a small factor, and in many circumstances can be assumed to be the equilibrium value. When this is so

$$p_n/p_{p0} = \exp(-eV_j/\kappa T) = n_p/n_{n0} \qquad (3.14)$$

If we use the semiconductor equation to provide an expression for the minority-carrier density, and substitute this in eqn (3.4), we can obtain

$$p_{n0}/p_{p0} = \exp(-eV_b/\kappa T) = n_{p0}/n_{n0}, \qquad (3.15)$$

The applied voltage $V_{ext} = V_j - V_b$, so that, combining eqns (3.14) and (3.15)

$$p_n/p_{n0} = \exp(-eV_{ext}/\kappa T) = n_p/n_{p0}. \qquad (3.16)$$

Put into words: the minority-carrier density on one side of a p-n junction is physically related to the majority-carrier density on the other side by a term containing the total potential difference between the two sides. But if we like we may relate the perturbed carrier density to the equilibrium carrier density by an exponential term containing the externally applied voltage. This is often very convenient, as the applied voltage is easy to measure, while V_j is not.

Fig.3.5 shows the relations pictorially. When the density is plotted on a logarithmic scale, then both the minority-carrier densities will be moved the same distance on the graph away from the equilibrium condition. An important point is that more minority carriers come from the more heavily doped side - this is exploited in the base-emitter junction of a

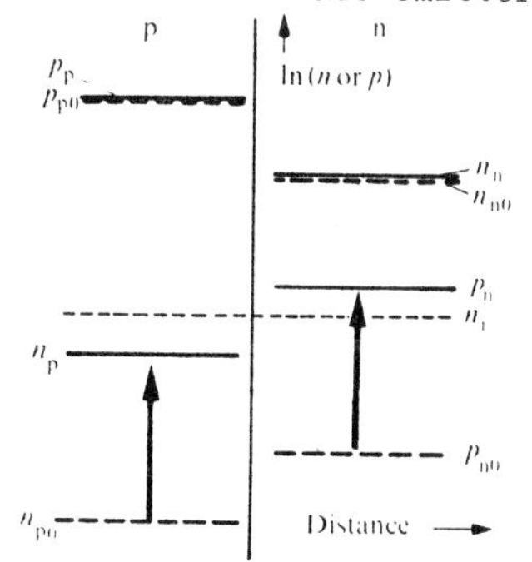

Fig.3.5 Carrier densities across a p-n junction with forward bias. On a ln(density)scale, n_i comes midway between the majority carrier densities, and both minority-carrier densities are raised by the distance $\ln\{\exp(eV_{ext}/\kappa T)\}$ when a voltage V_{ext} is applied.

bipolar transistor discussed on pp

VOLTAGE-CURRENT RELATION FOR A p-n JUNCTION DIODE

By combining the results of the previous section with those of (1.17), we can describe the flow of carriers across and away from the depletion layer in a p-n junction diode. We consider a diode which is much thicker than the diffusion lengths L_e or L_h (Fig.3.6). The continuity equation (1.26) for the excess electron density n' is

$$D_e(d^2n'/dx^2) - n'\tau = 0 . \tag{3.17}$$

This is simpler than eqn (1.26), because the electric field is low enough to ignore if the currents are not too large, the equilibrium generation and recombination terms have cancelled out, and the total rate of change of density is zero in a steady situation. In the p-side we can write the excess density as a function of the distance x from the edge of the depletion layer, $n'_p(x)$.

From eqn (3.16), the excess density at $x = 0$ is

$$n'_p(0) = n_{p0}\{\exp(eV_{ext}/\kappa T) - 1\}$$

The excess density falls to zero as $x/L_e \gg 1$, so the solution to eqn (3.17) is

$$n'_p(x) = n'_p(0)\exp(-x/L_e). \tag{3.18}$$

From eqn (3.16), the excess density at $x = 0$ is

$$n'_p(0) = n_{p0}\{\exp(eV_{ext}/\kappa T) - 1\}.$$

The current density of the diffusing electrons is

$$j_e(x) = eD_e dn'_p(x)/dx,$$

and at the edge of the depletion layer ($x = 0$) it is

$$j_e(0) = (eD_e n_{p0}/L_e)\{\exp(eV_{ext}/\kappa T) - 1\}.$$

A similar analysis for holes injected into and flowing in the n-side gives a similar expression for the current density

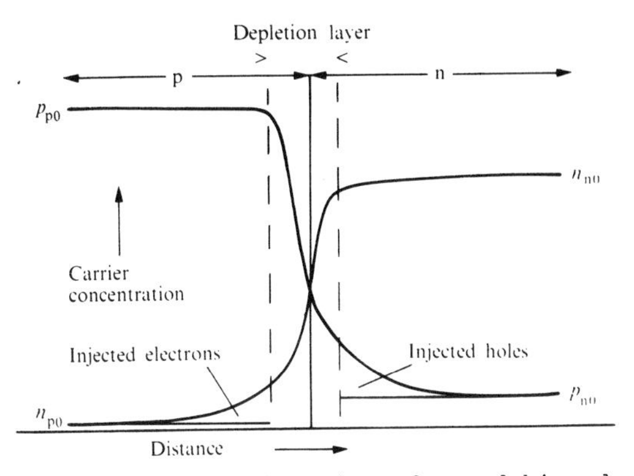

Fig.3.6. Carrier concentrations in a forward-biased p-n junction diode.

carried across the depletion layer by holes. The total current density j is the sum of the two contributions

$$j = j_e(0) + j_n(0)$$
$$= e\{(D_e n_{p0}/L_e) + (D_h p_{n0}/L_h)\}\{\exp(eV_{ext}/\kappa T) - 1\} \quad (3.19)$$

The first part of the equation is a constant for a given diode, and can be written as j_r, so the whole equation then becomes

$$j = j_r\{\exp(eV_{ext}/\kappa T) - 1\}.$$

For a diode of area A, the total current I will be jA, and if we define $I_r = Aj_r$, the d.c. voltage-current relation for a p-n junction diode is

$$I = I_r\ \{\exp(eV_{ext}/\kappa T) - 1\} \quad (3.20)$$

(check that the equation behaves in a suitable way to fit Fig.3.1). The expression for I_r can be written out in terms of the doping density rather than the minority-carrier density.

$$I_r = Aen_i^2\{(D_e/L_e N_d) + (D_h/L_h N_a)\} \quad (3.21)$$

For many purposes a p-n junction is required to have I_r small. Inspection of eqn (3.21) shows what is needed to achieve this - in particular n_i can be made small by choosing a material with a high band gap, and by avoiding high temperatures, and N_a and N_d should both be large.

The theory leading to eqn (3.20) is an accurate description of some p-n junction diodes for moderate forward currents and reverse voltages. For other diodes other effects are important in addition to those included in our analysis. For instance, the resistance of thick semiconductor layers may increase the total voltage of the diode for a given current; the depletion layer may be wide enough for a noticeable number of electron-hole pairs to be generated within it; the minority-carrier densities may rise to approach the majority-carrier densities. Theories including some or all of these more advanced effects can be found elsewhere (see Sze 1969, van der Zeil 1968), but the Boltzmann equation, minority-carrier injection, and diffusion will always form a basis for understanding the physical processes occuring in p-n junction diodes.

The small signal differential resistance r of a diode can be obtained by differentiating eqn (3.20). It has a simple form when $\exp(eV_{ext}/\kappa T) \gg 1$:

$$r = \frac{dV}{dI} = \left(\frac{dI}{dV}\right)^{-1} = \frac{\kappa T}{eI} \tag{3.22}$$

At 300 K, $\kappa T/e = 1/40$ V, so that if I is in amperes the value of r in ohms is $1/40I$.

This is a useful formula for describing the input impedance of a bipolar transistor.

DIFFUSION CAPACITANCE

When a p-n junction diode is passing current, there are more electrons and holes in it than when the current is zero. These charges have to be supplied, and as a result the diode acts as a capacitor with a capacitance C_d, additional to $C(V_j)$.

Define C_d as dQ/dV. For this problem it is convenient to think of the relation of the extra charge Q to the diode current

I, so we write

$$C_d = dQ/dV = (dQ/dI)(dI/dV).$$

The value of Q is found by integrating the excess charge density described by eqn (3.18). For the electrons in the p-side,

$$Q_e = \int_0^\infty eAn_p(x)dx = eAn_{p0}\{\exp(eV/\kappa T)-1\}\int_0^\infty \exp(-x/L_e)dx \ .$$

Hence

$$Q_e = eAn_{p0}L_e\{\exp(eV/\kappa T) - 1\}.$$

There will be a similar expression for the excess holes in the n-side, so that the total excess charge is

$$Q = eA\{\exp(eV/\kappa T)-1\}(n_{p0}L_e + p_{n0}L_h)$$

or, for compactness,

$$Q = Q_0\{\exp(eV/\kappa T) - 1\}.$$

Thus, from eqn (3.20), we can write

$$Q = Q_0 . I/I_r, \quad dQ/dI = Q_0/I_r \ .$$

We already have a form for dI/dV from eqn (3.22) so that

$$C_d = (eI/\kappa T)(Q_0/I_r) \tag{3.23}$$

The diffusion capacitance increases as the forward current increases, becoming larger than the depletion-layer capacitance for all reasonable forward biases. It is always associated with the diode differential resistance r, so the forward-biased diode is inevitably lossy, and cannot be used as a way of making a good capacitor.

BIPOLAR TRANSISTORS

The bipolar junction transistor, which rapidly supplanted the earlier point-contact transistor, has been the financial foundation of semiconductor electronics. From now on, when the term 'transistor' is used without qualification, then a bipolar junction transistor should be understood.

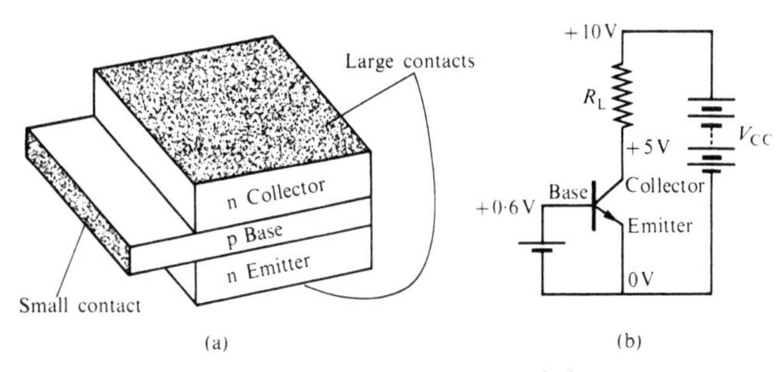

Fig.3.7. An n-p-n bipolar transistor: (a) schematic construction; (b) conventional symbol and circuit showing 'normal bias'. V_{cc} is the supply to the collector and load.

Construction and low-frequency characteristics

A simple version of the shape of the active regions of a transistor is shown in Fig.3.7(a). The three layers of semiconductor which constitute the emitter, base, and collector may be n-p-n or p-n-p, giving two alternative arrangements. The two kinds of transistor complement each other's function in electronic circuits, the signs of currents and voltages being reversed. The physical explanations required for the two kinds are the same if holes be read for electrons, p for n, and so on. The main carrier flow is from the emitter through the base and out through the collector, so the emitter and collector have large contacts. The base is thin - as thin as can be made (0·2 μm or less), and it has a contact at the side for a third, controlling connection.

In use the base-emitter junction is forward-biased (as if it were a diode) and the base-collector junction is reverse-biased by 1 - 20 V (Fig.3.7(b)). Fig.3.8(a) and (b) show the collector current as a function of collector-emitter voltage, for various values of the base current for a Si n-p-n transistor and a Ge p-n-p transistor. Notice that the current increases very rapidly as V_{CE} goes from 0 V to about 0·5 V, and then increases

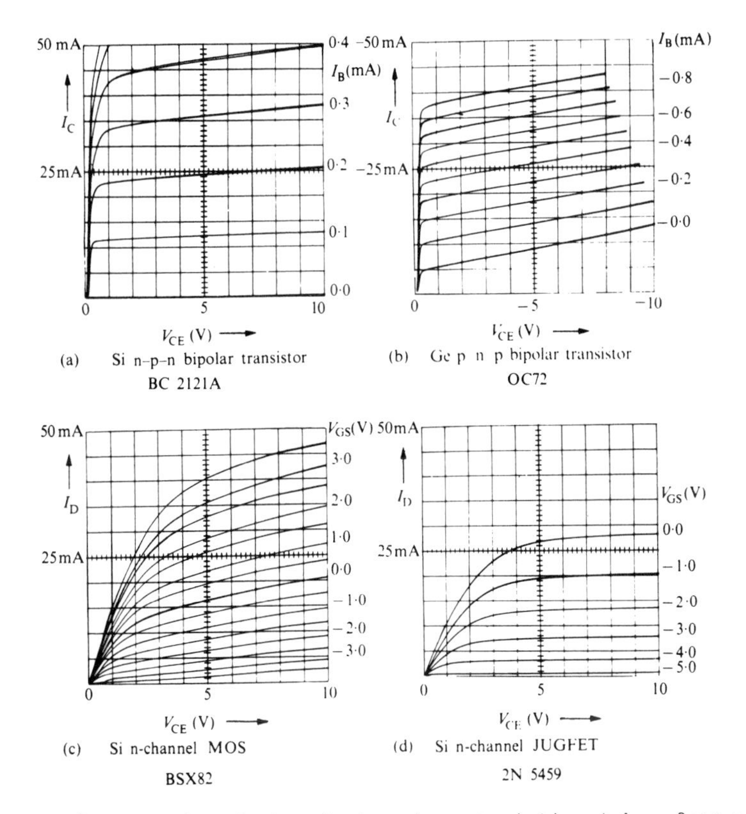

Fig.3.8. Semiconductor device characteristics taken from a Tektronix curve tracer; (a) the wide spacing of the curves shows that dI_C/dI_B is large: (b) the p-n-p device has the sign of I_C and V_{CE} reversed; the tips of the curves mark out a 'load line'; (c) V_{GS} can be positive or negative; the traces near the origin show some instrumental kinks; (d) the JUGFET is more closely related to the MOS than to the bipolar transistor.

much less rapidly. For the Si transistor, the collector current is very close to zero when the base current is zero, but this is not so for the Ge transistor. An important parameter is (dI_C/dI_B) for constant V_{CE} (often called β). The steps of I_B in the two sets of curves are the same, so the larger spacing of the curves for the Si device implies a higher β (what is its numerical value?)

With a suitable load resistor R_L the collector voltage can vary by much more than the changes in the base-emitter voltage. Consequently small changes of current and voltage can be amplified into much larger ones, a function with many applications.

The limitations of a transistor in amplying high-frequency signals cannot be assessed from curves such as those in Fig(3.8). The high-frequency characteristics are discussed on p.77.

The maximum current that can be passed through a transistor is limited by the need to keep the transistor from overheating. High-power transistors need good heat sinks. The maximum voltage that can be applied between collector and base may be limited by avalanche breakdown in the collector-base depletion layer, or by the depletion layer extending right across the base and reaching the emitter, this process being called 'punch through'. Maximum values of V_{CE} may be from 5 V to 1500 V.

Physical principles of bipolar transistors

Three physical principles make a useful first description of the way a bipolar transistor works.

(a) The externally applied voltages control the ratio of the carrier densities on the two sides of each depletion layer (see p.56).

(b) Away from the depletion layer the regions are neutral.

(c) Hence carriers move away from the junctions after injection by diffusion and the action of any fields built in by variable doping. The applied voltages

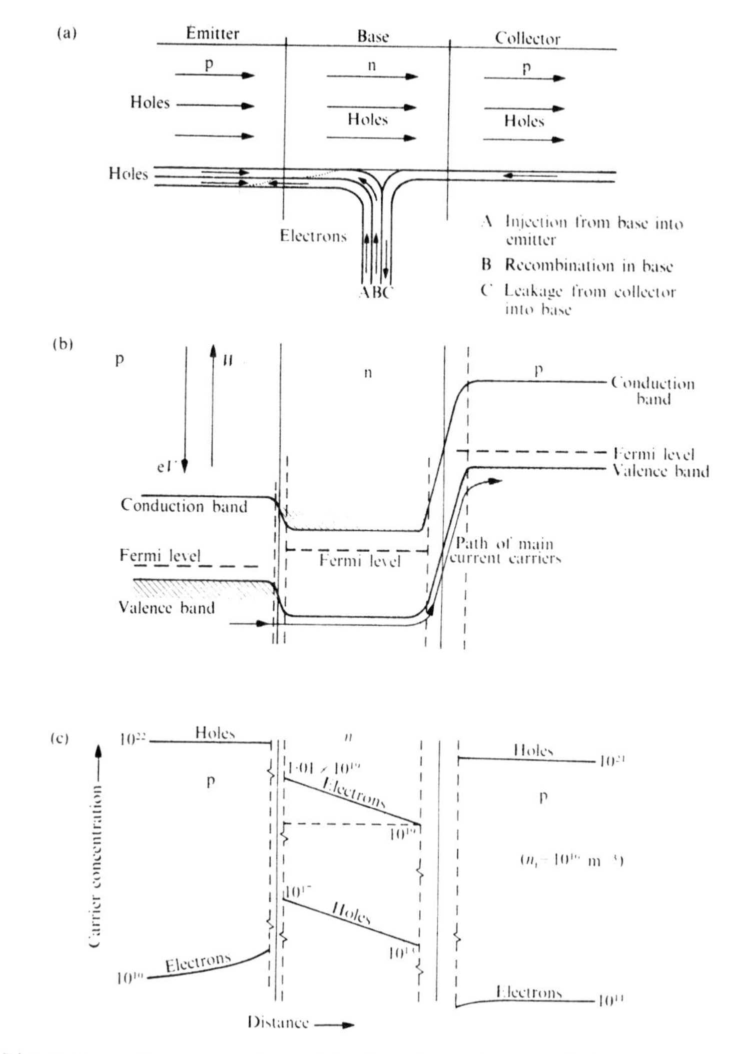

Fig.3.9. Processes in a bipolar transistor; (a) fluxes of holes and electrons; (b) band structure and potentials; (c) hole and electron concentrations.

control the carrier flow solely through the injection process.

Fig.3.9 shows these ideas for a p-n-p transistor from three aspects: the variation along a line from emitter to collector of the potential and the band structure, the fluxes of particles, and the carrier densities.

In Fig.3.9(a), the width that a stream is drawn indicates the current it carries, and the arrows point in the direction of particle flow not in the direction of the conventional electrical current. The largest current is of holes injected into the base. A fraction f_1 of these holes combine with electrons in the base, but most diffuse to the collector-base depletion layer where they are whisked away into the collector by the high field in the depletion layer.

Injection always works both ways, though to different extents, and there is a current of electrons injected back into the emitter, which is proportional to the forward current of holes I_{Eh} (let it be $f_2 I_{Eh}$). In useful transistors f_1 and f_2 are arranged to be small.

The total emitter current is $I_{Eh}(1 + f_2)$, while the total base current is $I_{Eh}(f_1 + f_2)$, so their ratio is

$$(1 + f_2)/(f_1 + f_2) \sim 1/(f_1 + f_2).$$

This is the common-emitter current gain β of the transistor.

From the emitter there is a fairly constant leakage current to the base of electrons. In Si transistors this is often so small that it can be ignored.

The mention of both electrons and holes in this discussion justifies the name <u>bipolar</u> transistor, by comparison with some field-effect transistors, (see p.85) where only the majority carriers appear to play an important part. In Fig.3.9(b), the relative positions of the Fermi level show the external bias voltages. In a p-n-p transistor the base is slightly negative relative to the emitter, and the collector is very negative. There is no field in the base, which is correct for a uniformly doped base. The results of non-uniform doping in the base are discussed on page 77. Notice how the majority carriers in the base are in a potential well, with a barrier on each side of them. This explains why the minority carriers carry most of the current, because the majority carriers are, to a useful extent, trapped in the base.

The depletion region between emitter and base is thin because the potential across it is small. The depletion region between base and collector has a much higher voltage across it, and its thickness has to be taken into account (see p.51).

The carrier densities in Fig.3.9(c) are difficult to draw so that all the points are clearly demonstrated, as both very large and very small densities are important. In the base the density of the majority carriers (electrons) makes a line of the same slope as that for the minority carriers, so the region is everywhere neutral. The electrons ought therefore to diffuse, but since they are trapped, a small field builds up which is just sufficient to balance the diffusion current, but, except at high currents, is too small to have a noticeable effect on the far fewer holes. The injection of carriers back into the emitter can be seen, but the recombination of carriers in the base is too small an effect to be evident, though if it were shown the density in the base would fall a little more steeply near the emitter than near the collector.

The Ebers-Moll model for a bipolar transistor

The ideas sketched in Fig.3.9 can be put down in equations, and we shall do this for a simple case where all three regions are so short that recombination can be completely neglected except at the contacts (Fig.3.10). Our results will apply to any combination of bias voltages - large or small, positive or negative - and lead to the Ebers-Moll equations for a bipolar transistor, which describe d.c. behaviour in a very compact way. Were we to take recombination in the bulk into account, the mathematics would become more involved, but the form of the final equation would be unaltered.

The first of the three principles stated at the beginning of the previous section may be expressed using the notation of Fig.3.10, for the base-emitter junction,

$$\frac{p_{BE}}{p_{BO}} = \exp\left(\frac{eV_{EB}}{\kappa T}\right) = \frac{n_{EB}}{n_{EO}} , \tag{3.24}$$

and for the base-collector junction,

$$\frac{p_{BC}}{p_{BO}} = \exp\left(\frac{eV_{CB}}{\kappa T}\right) = \frac{n_{CB}}{n_{CO}} . \tag{3.25}$$

For a transistor of area A without any field built into the base, the second and third principles allow us to describe the total current as made up of diffusion currents.
For the base, the hole current is

$$I_{Bp} = AeD_h(p_{BE} - p_{BC})/W_B . \tag{3.26}$$

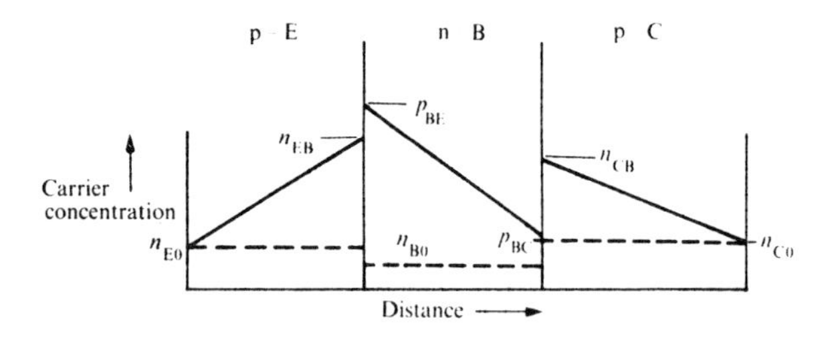

Fig.3.10. Model for the Ebers-Moll analysis. Recombination occurs only at contacts, so the graphs of n or p against distance are straight. Read a symbol like n_{CB} as 'the concentration of electrons at the side of the collector nearer the base', and a symbol like p_{EO} as 'the concentration of holes in the emitter in equilibrium'.

For the emitter and collector, the electron-diffusion current is

$$I_{En} = AeD_e(n_{EB} - n_{EO})/W_E \tag{3.27}$$

$$I_{Cn} = AeD_e(n_{CB} - n_{CO})/W_C.$$

Because volume recombination is absent from our model, the hole current in the base (eqn.3.26) must also be the hole current in the emitter and collector, so we know the total current in each region. The total base current is fixed by

$$I_E + I_B + I_C = 0, \tag{3.28}$$

where we are defining a positive current as flowing into each terminal. This is the end of the physics - the rest is algebra to put the equations into revealing forms.

It is useful to define three quantities which are characteristic of the three parts of the transistor,

$$E = AeD_e\ n_{EO}/W_E,$$

$$B = AeD_h\ p_{BO}/W_B,$$

$$C = AeD_e\ n_{CO}/W_C\ .$$

The total emitter and collector currents, I_E and I_C, may now be written out, using (3.24) and (3.25) for the densities in (3.26) and (3.27),

$$I_E = I_{En} + I_{Ep} = E\{\exp\left(\frac{eV_{EB}}{\kappa T}\right) - 1\} + B\{\exp\left(\frac{eV_{EB}}{\kappa T}\right) - \exp\left(\frac{eV_{CB}}{\kappa T}\right)\},$$

$$I_C = I_{Cn} + I_{Cp} = C\{\exp\left(\frac{eV_{CB}}{\kappa T}\right) - 1\} + B\{\exp\left(\frac{eV_{CB}}{\kappa T}\right) - \exp\left(\frac{eV_{EB}}{\kappa T}\right)\} \tag{3.29}$$

In these equations the transistor looks like a voltage-controlled device, because if the voltages are known the currents can easily be found. This approach is helpful in understanding why the transistor works, but a transformation can put the equations into another form which has been found more suitable for circuit work.

First we re-cast each expression in (3.29) into the form of diode equations:

$$I_E = (E+B)\left\{\exp\left(\frac{eV_{EB}}{\kappa T}\right)-1\right\} - \frac{B}{C+B}(C+B)\left\{\exp\left(\frac{eV_{CB}}{\kappa T}\right)-1\right\}$$
$$I_C = (C+B)\left\{\exp\left(\frac{eV_{CB}}{\kappa T}\right)-1\right\} - \frac{B}{E+B}(E+B)\left\{\exp\left(\frac{eV_{EB}}{\kappa T}\right)-1\right\}. \tag{3.30}$$

Next we introduce symbols which relate to directly measureable properties of the transistor in place of E, B, and C,

$$I_{EBo} = (E + B), \quad I_{CBo} = (C + B),$$

$$\alpha_i = B/(C + B), \; \alpha_n = B/(E + B).$$

Clearly, $\alpha_n I_{EBo} = \alpha_i I_{CBo}$. The Ebers-Moll equations now take the form

$$I_E = I_{EBo}\left\{\exp\left(\frac{eV_{EB}}{\kappa T}\right)-1\right\} - \alpha_i I_{CBo}\left\{\exp\left(\frac{eV_{CB}}{\kappa T}\right)-1\right\};$$
$$I_C = I_{CBo}\left\{\exp\left(\frac{eV_{CB}}{\kappa T}\right)-1\right\} - \alpha_n I_{EBo}\left\{\exp\left(\frac{eV_{EB}}{\kappa T}\right)-1\right\}. \tag{3.31}$$

To obtain a value for I_{EBo}, we measure I_E when the emitter-base junction is reverse-biased so that $\exp(eV_{EB}/\kappa T)$ is very small, and with the collector open circuit, i.e. with $I_C = 0$ (check the relation between I_{EBo} and the measured value of I_E - they are not the same). A value of I_{CBo} can be found in an equivalent way. To find α_n (α_{normal}) we make V_{CB} very large and negative, so that

$$I_C = I_{CBo} - \alpha_n I_{EBo}\{\exp(eV_{EB}/\kappa T)-1\} = I_{CBo} - \alpha_n(I_E - \alpha_i I_{CBo}).$$

Then $\alpha_n = -dI_C/dI_E$. The bias condition is the normal one for the ordinary use of the transistor, and we expect α_n to be a little less than unity. If the function of the emitter and collector are interchanged, the transistor will still work as a transistor, though probably less well. The value of α_i ($\alpha_{inverted}$) can be found by making V_{EB} large and negative, and measuring $-dI_E/dI_C$.

We can now build up an equivalent circuit for the d.c.operation of a transistor from eqn (3.31). The term $I_{EBo}\{\exp(eV_{EB}/\kappa T)-1\}$ could be the description of a junction diode with $I_r = I_{EBo}$. We can represent it on the equivalent circuit by a diode symbol, with I_{EBo} written beside it. The diode goes between emitter and base, because the current is part of the emitter current, and the diode voltage is V_{EB}.

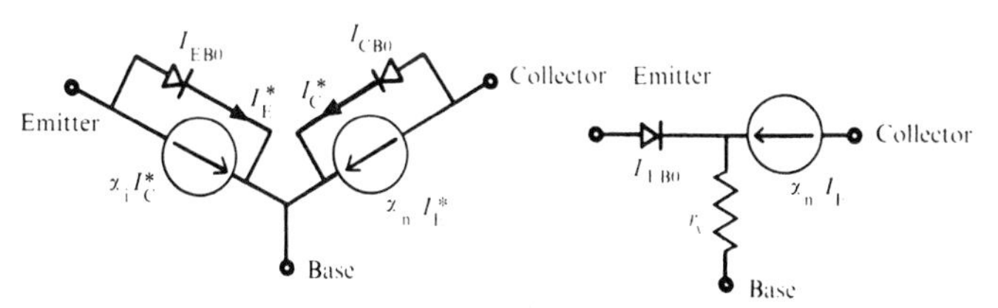

Fig 3.11. Ebers-Moll equivalent circuit: (a) for any bias condition: (b) a simplified circuit which applies when normal working bias is applied, and r_x is added.

The term $-\alpha_i I_{CBo}\{\exp(eV_{CB}/\kappa T)-1\}$ requires a current generator which states that the emitter current has a contribution $-\alpha_i I_C^*$ (I_C^* is defined in Fig.3.11(a)). The components in the collector branch are read in the same way from the equation for I_C. The process of building up Fig.3.11(a) by reading the terms in the equations should be noted, as it is a technique characteristic of the physics of electronic devices.

When the normal set of bias voltages are applied to a transistor (V_{CB} large and negative, V_{EB} small and positive for our p-n-p transistor), Fig.3.11(a) simplifies to Fig.3.11(b), if we may neglect I_{CBo}. The resistor in the base branch is not part of the formal Ebers-Moll model, but has been added to the circuit to allow for the resistance of the semiconductor away from the active region of the base. We shall use Fig.3.11(b) as the basis for our analysis of the noise properties of a transistor, and could use it as a model for the small-signal properties of a transistor by representing the diode by its slope resistance. Instead of so doing we shall develop the hybrid-π model(in the next section) as a more flexible small-signal equivalent circuit.

The Ebers-Moll equations go a long way to giving a good description of the d.c. characteristics of bipolar transistors, but they omit an important effect. As V_{CB} increases the collector-base depletion layer expands, so that the remaining neutral base width becomes less and less. As a result the collector current increases with V_{CE}, as can be seen in Fig.3.7(a) and (b). If the current were constant the slope resistance would be infinite, and would lead to very different circuit effects, so the change of base width or base-width modulation is important.

Hybrid-π equivalent circuit for bipolar transistors

Transistors are often used to amplify small signals, so it is important to have some model to describe the way a transistor operates in an amplifying circuit. We assume that the transistor has suitable steady bias voltages and currents, and shall expect the values of the components of the equivalent circuit to depend on the quiescent conditions. We wish our model to apply over a wide range of frequency with reasonable accuracy.

Many small-signal equivalent circuits have been suggested for transistors, and indeed they can be transformed one into another. Here we examine the hybrid-π model, because its elements can be related to the physical processes occuring in the transistor. We start with the idea that the density gradient of excess carriers in the base is constant (or nearly so), and that the current is proportional to the density gradient (Fig.3.12(a),

$$I_C = -AD_e e(dn/dx).$$

The idea holds if the carriers have time to distribute themselves

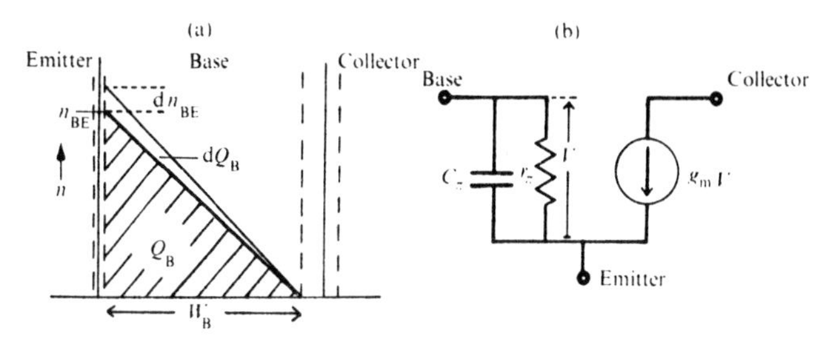

Fig.3.12. Hybrid-π small-signal equivalent circuit: (a) the excess charge in the base is Q_B, and the density gradient $dn/dx = n_{BE}/W_B$: (b) the first three elements of the hybrid-π equivalent circuit.

properly - when this condition is not met, the frequency of operation is so high that the gain of the transistor has become low.

The density falls from $n_{BE} = n_{BO}\exp(eV_{EB}/\kappa T)$ on the emitter side of the base to a low value on the collector side, so the density gradient is

$$\frac{dn}{dx} = -(n_{BO}/W_B)\exp(eV_{EB}/\kappa T) ,$$

where we use the same notation as in the previous section. Thus

$$I_C = (AeD_e n_{BO}/W_B)\exp(eV_{EB}/\kappa T) .$$

$$\left.\frac{dI_C}{dV_{EB}}\right|_{V_{CE}} = \frac{e}{\kappa T} I_C$$

or

$$dI_C = (e/\kappa T)I_C\, dV_{EB} . \qquad (3.32)$$

Eqn (3.32) can be put in terms of a transfer conductance $g_m = (e/\kappa T)I_C$, as we have a change in the voltage between two terminals controlling a current to the third terminal.

Then $dI_C = g_m\, dV_{EB}$, and we have the first element of the hybrid-π equivalent circuit (Fig.3.12(b). A particular convenience is that the value of g_m depends only on quantities that are easy to find. The steady collector current is related to the total excess charge in the base Q_B,

$$I_C = eAD_e\, n_{BE}/W_B,$$

$$Q_B = \tfrac{1}{2}eAn_{BE}\, W_B\,. \qquad (3.33)$$

Thus

$$I_C = \frac{Q_B}{(W_B^2/2D_e)}\,.$$

Since current is a rate of flow of charge, $W_B/2D_e$ is the time taken for Q_B to flow through the base, and is given the symbol τ_t, the minority-carrier transit time.

The change in the base current when the base-emitter voltage is altered from its quiescent value has two components - steady and transient. The steady part supplies the change in the loss processes in the base, namely, the recombination of carriers in the base and the injection of carriers back into the emitter. If the excess majority-carrier lifetime in the base due to these causes is τ_B, the steady current needed to replenish the carrier density is

$$I_B = Q_B/\tau_B\,.$$

Since Q_B is related to n_{BE} (eqn (3.33)) and hence to V_{EB}, we can find how I_B is related to V_{EB},

$$\frac{dI_B}{dV_{EB}} = \frac{eAW_B}{2\tau_B}\cdot\frac{d}{dV_{EB}}\left\{n_{BO}\exp\left(\frac{eV_{EB}}{\kappa T}\right)\right\}.$$

Thus we now have an expression for g_π, the conductance between base and emitter.

$$g_\pi = \frac{e^2AW_B}{2\tau_B}\cdot\frac{n_{BO}}{\kappa T}\exp\left(\frac{eV_{EB}}{\kappa T}\right) = \frac{1}{r_\pi}\,, \qquad (3.34)$$

where r_π, the resistance between base and emitter, is used as an alternative notation to $1/g_\pi$.

A little substitution (try it) shows that

$$g_\pi = g_m \cdot \tau_t/\tau_B \,. \tag{3.35}$$

Since τ_t is always much less than τ_B, $g_\pi \ll g_m$, and only small changes in I_B are needed to produce large changes in I_C. The ratio $(\tau_t/\tau_B)^{-1}$ is β, the common-emitter current gain.

The transient part of the base current is needed to bring Q_B up to its new value

$$\frac{dQ_B}{dV_{EB}} = \frac{dn_{BE}}{dV_{BE}} \cdot \frac{eAW}{2} = \frac{e}{\kappa T} Q_B = \frac{e}{\kappa T}(I_C \tau_t).$$

The expression dQ_B/dV_{EB} describes a capacitance C_π, in the equivalent circuit, where

$$C_\pi = (e/\kappa T)(I_C \tau_t).$$

In Fig 3.12(b) we have the three main elements of the hybrid-π equivalent circuit, and these alone would give a useful approximate description of the variation of the performance of the transistor with frequency.

The next group of components depends on base-width modulation. This is the reduction of W_B as V_{CB} increases (Fig 3.13(a)), which has three important consequences.

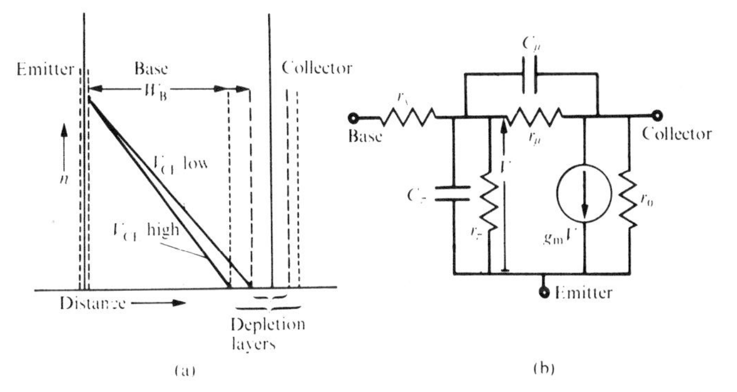

Fig.3.13. Hybrid-π small-signal equivalent circuit: (a) an increase in the collector-emitter voltage difference reduces the base width and increases dn/dx: (b) the complete hybrid-π equivalent circuit.

There is an increase in dn/dx, and hence I_C for a given V_{BE}. On the equivalent circuit we represent this by a conductance g_O between emitter and collector. A value for g_O is best obtained from curves like those in Fig 3.8, rather than by calculation.

Another result of base width modulation is that Q_B is reduced, so a transient reverse current is called for. A capacitance C_μ between collector and base provides this. The reduced Q_B requires reduced replenishment, so the total steady base current should reduce as V_{CE} increases. A conductance g_μ between collector and base does this job. Notice that C_μ and g_μ are the first components we have discussed that allow the feedback of a signal from collector to base. In general, the smaller they are, the better. Like g_O, g_μ and C_μ are better determined from measurement rather than calculation.

The two capacitances C_μ and C_π are physically related to the diffusion capacitance analysed on p 54. The other kind of p-n junction capacitance - depletion-layer capacitance - is also present at the two junctions. In the equivalent circuit the depletion-layer capacitances may be added on to C_μ and C_π.

Current flowing to the active region of the base between collector and emitter has first to pass through a volume of lightly doped semiconductor, which may have enough resistance to matter. It is found to be worthwhile including this on the equivalent circuit as an extra resistance r_x, which is independent of the bias conditions, having a value between 10 Ω and 100 Ω. The voltage which controls the current generator g_m is now not the voltage between the emitter and base terminals, but the voltage relative to the emitter of an internal node on the equivalent circuit, which represents the signal reaching the p-n junction.

The full hybrid-π equivalent circuit (Fig 3.13(b)) should give a close description of the way a transistor with one set of bias conditions handles small signals over its full range of working frequencies.

High-frequency limits for a bipolar transistor

The amplification available from a transistor circuit falls at high frequencies. A useful figure of merit is the frequency (f_t) where the current gain in the common-emitter circuit with zero load impedance has fallen to unity. The value of f_t can be related to the total delay of minority carriers passing from emitter to collector. Delays which may be important can occur in changing the charge at the emitter-base depletion layer, in diffusing across the base, and in changing the charge at the collector.

In some transistors the delay τ_t in diffusing across the base, is the longest. We have seen that $\tau_t = W_B^2/2D$, and hence expect

$$f_t = 1/(2\pi\tau_t) = D/(\pi W_B^2) .$$

This conclusion can be confirmed by looking at a simple version of the hybrid-π equivalent circuit which omits C_μ and g_μ. The only frequency-dependent element is C_π, which in parallel with g_π gives a time constant τ_t (check this using the expressions for C_π and g_π - remember that the time constant of a capacitance and a conductance is C/g).

For transistors made using the planar process (see p. 129) a doping density gradient in the base builds in a field which hastens the minority carriers on their way, so that τ_t is less than $W_B^2/2D$, and f_t is raised. On the other hand, the light doping in the collectors of these devices results in a greater need to build up the charge density in the collector, and a correspondingly greater delay in the collector. To some extent W_B is under the user's control. By using a high value of V_{CB}, the depletion layer can be expanded and W_B reduced, thereby raising f_t.

Transistors are available with values of f_t up to 10 GHz; at these frequencies the connecting leads have to be designed to be part of the microwave circuit - merely making them short is not enough.

The charge-control model

The charge-control model for bipolar transistors is used to describe their action in pulse circuits. Unlike the hybrid-π equivalent circuit, it does not describe small variations about a specified bias point, but is suitable for situations where the transistor changes from being completely on to completely off, as in many digital applications.

In the charge-control model, attention is focused on the excess charge Q_B in the base of the transistor, and the equivalent circuit is built up of components whose value is controlled by Q_B. The need to build up and replenish Q_B is expressed by some of the equivalent-circuit components.

On page 74 we saw that the relations between Q_B, I_C, and I_B were

$$I_C = Q_B/\tau_t \,, \qquad I_B = Q_B/\tau_B \,. \tag{3.36}$$

These relations will apply both to transistors with a uniformly doped base, when $\tau_t = W_B^2/2D$, and to transistors with a density gradient of doping atoms in the base. If we write $\tau_B = \beta\tau_t$ we can see equivalent circuit components corresponding to eqn (3.36) in Fig.3.14(a). The symbol labelled Q_B is not a capacitor, as the voltage across it does not increase as the charge increases; it is sometimes called a *storance*. The full set of equations corresponding to Fig.3.14(a) is

$$\begin{aligned} I_B &= Q_B/\beta\tau_t + dQ_B/dt, \\ I_C &= Q_B/\tau_t \,, \\ I_E + I_C + I_B &= 0 \,. \end{aligned} \tag{3.37}$$

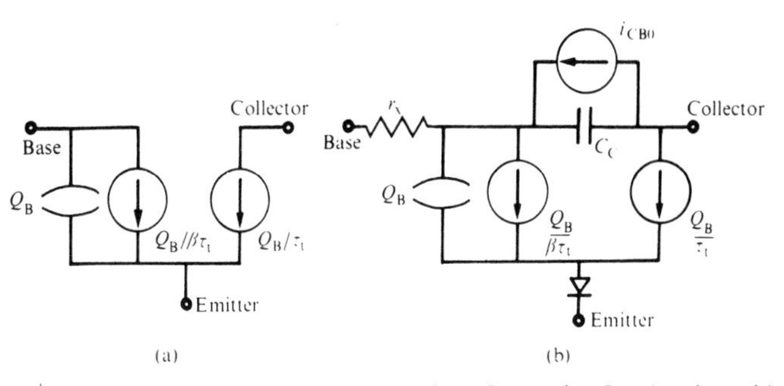

Fig.3.14. Charge control large-signal equivalent circuit: (a) the three circuit elements shown form the simplest charge control equivalent circuit; Q_B is the excess charge (majority or minority) stored in the base: (b) an extended equivalent circuit.

This is the basic charge-control model, and describes both a.c. and d.c; but to be useful it needs further development.

Fig.3.14(b) is a full model for conditions where the transistor is not saturated, i.e. where the collector-base junction is reverse-biased, and all minority carriers reaching it are swept across the depletion layer. The new components of the equivalent circuit are:

(a) A current generator representing the leakage current from collector to base, taken to be independent of V_{CB}.

(b) The depletion-layer capacitance C_C for the collector-base junction. To make the model simple we take an average value for C_C, even though as V_{CB} increases, C_C will fall.

(c) The depletion-layer capacitance C_E for the emitter-base junction. We can take an average value, or, noting that V_{BE} varies only little, realize that the charge in C_E is unchanging, and ignore it.

(d) A diode in the emitter lead. It represents the voltage between emitter and base, and can be as exact a description of a diode as is required, but has no stored charge or delays - it is an 'instantaneous diode'.

(e) A resistance r_x, representing the passive base region. The equations corresponding to Fig.3.14(b) are

$$I_B = Q_B/\beta\tau_t + dQ_B/dt + dQ(C_C)/dt - I_{CBo} = 0 ,$$
$$I_C = Q_B/\tau_t - dQ(C_C)/dt + I_{CBo} = 0 , \qquad (3.38)$$

$$dQ(C_C)/dt = -C_C dV_{CB}/dt .$$

Notice that C_E has been ignored, and r_x and the diode appear only when terminal voltages are being calculated.

When the base current is larger than is needed to maintain the maximum current that the circuit can attain (V_{CC}/R_L Fig.3.7(b)), the current and the base charge are no longer related by $I_C = Q_B/\tau_t$, a new relation is needed. The base charge has to be divided between Q_B which still relates to I_C as before, and Q_{Bs}, the saturated base charge (Fig.3.14(c)), which does not contribute to the density gradient though it is stored in the base. The saturated base charge decays at a rate $dQ_{Bs}/dt=Q_{Bs}/\tau_s$, where τ_s is comparable to $\beta\tau_t$, though it may well be somewhat shorter.

If, in a circuit, V_{CC} were suddenly to increase or R_L were to decrease, then Q_{Bs} could be redistributed to permit a rapid increase in I_C without the need for an increase in the base current. Another situation in which Q_{Bs} has to be considered occurs when I_B falls to zero after a pulse. Until all the charge in the base has decayed (Q_{Bs} as well as Q_B) collector current can continue to flow. In practice this means that a transistor takes far longer to switch off than would otherwise be predicted.

The equations describing the charge-control model (Fig.3.15) for saturated conditions are

$$I_B = Q_B/\beta\tau_t + Q_{Bs}/\tau_s + dQ_{Bs}/dt,$$
$$I_C = Q_B/\tau_t . \qquad (3.39)$$

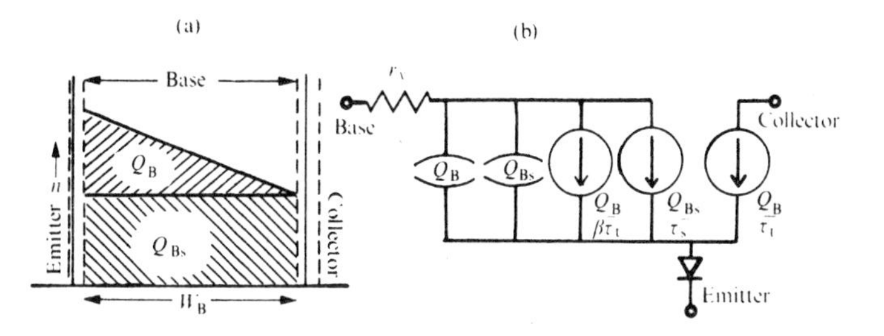

Fig. 3.15. The saturated charge-control model: (a) the base charge is made up of two parts, Q_B and Q_{Bs}, (b) the equivalent circuit is used while $I_B > I_C/\beta$.

The charge-control model is used by defining a starting situation and calculating how long a transistor takes to reach successive stages through a pulse, employing judicious approximations at each stage.

Two simplifications in the charge-control model presented here are the omission of any explicit description of base-width modulation (some average base width must be assumed) and the assumption that the density gradient is linear. With respect to the latter point the effect of a graded base doping introduces no error when the right values of τ_t are used, but a brief time (about $\frac{1}{3}\tau_t$) should be added when calculating the time taken for charges to reach their equilibrium positions in the base.

THYRISTORS

A number of semiconductor devices have been made with four or more layers of p and n alternately. We consider the thyristor as an example. It is made of four regions p-n-p-n, so there are three p-n junctions (Fig.3.16(a)). It can be switched ON by a current pulse to the gate, but once ON, can be switched OFF only by reducing the anode voltage to or below zero. It is used in low-frequency circuits to control power up to tens of kilowatts by selecting the instant of switching ON.

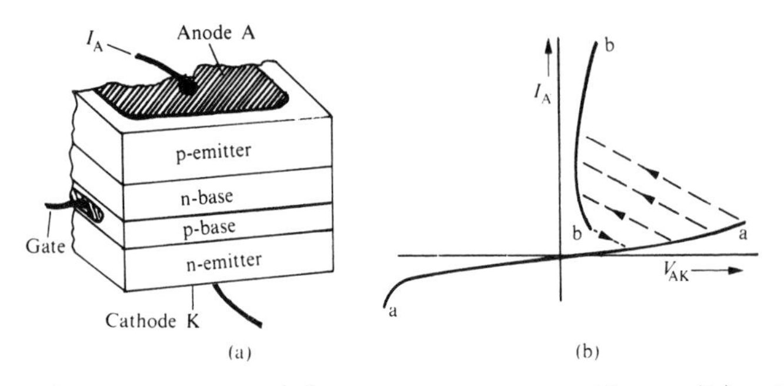

Fig.3.16. Thyristor: (a) schematic construction: (b) voltage-current characteristic: aa - OFF, bb - ON.

The anode-cathode voltage-current curves for a thyristor are shown in Fig.3.16(b). Along curve aa the thyristor is OFF, the anode may be positive or negative relative to the cathode, and any current flow is small. On curve bb, the thyristor is ON; the arrows show possible ways a transition between OFF and ON can occur. Notice that at some voltages, the current may be high or low, so we have to find some process inside the device which is different for high and low currents. Several non-linear processes have been suggested, but here we shall examine one only trap-controlled recombination, though not thereby implying that it is the full explanation in all cases.

Some impurities, Au in Si for instance, form energy levels in the forbidden band which are not near either band edge, so do not act as donors or acceptors. These levels are known as traps, or trapping states, and may 'capture' an electron (Fig.3.17(a)) which after a delay may escape back to the conduction band (Fig.3.17(b)) or fall to the valence band (Fig.3.17(c)). The last process can be also thought of as the further capture of a hole (Fig.3.17(d)).

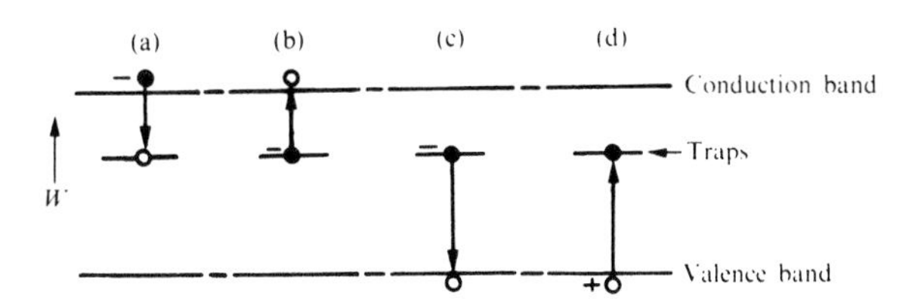

Fig.3.17. Carrier transitions between traps, and the valence and conduction bands: (a) an electron being trapped: (b) the electron becoming free again: (c) the trapped electron falling to the valence band: (d) the process in (c) may be represented in another way as the trapping of a hole.

Consider a situation where there is an excess electron density n', and the material has T traps per unit volume, of which a fraction f are full. Then the rate of loss of electrons from the conduction band is $n'(1 - f)TC$ where C is the rate constant for the capture. If process (b) on Fig.3.17 is so infrequent that it can be ignored, and process (c) can be described by a half-life τ, then for a steady fraction of the traps to stay filled,

$$n'(1 - f)TC = fT/\tau$$

or

$$f = \frac{n'C}{1/\tau + n'C} .$$

Notice that as n' becomes large $f \to 1$. The recombination rate can now be written

$$dn'/dt = n'TC/(1 + n'\tau C).$$

When n' is small, $dn'/dt \to n'TC$, so that the rate of recombination is proportional to n'. When n' is large, $dn'/dt \to T/\tau$, and hence is independent of n'. The average life of a carrier now becomes longer, and its diffusion distance becomes larger. The effect is observed in bipolar transistors, where the current gain for small emitter currents is low because few traps in the base are filled.

In a thyristor in the OFF condition, excess carrier densities are low, few traps are occupied, and minority-carrier lifetimes are short. In the ON state, excess carriers are high, and the traps are filled so that the minority-carrier lifetimes are much longer, demonstrating that trap-controlled recombination is a process that is non-linear in n'.

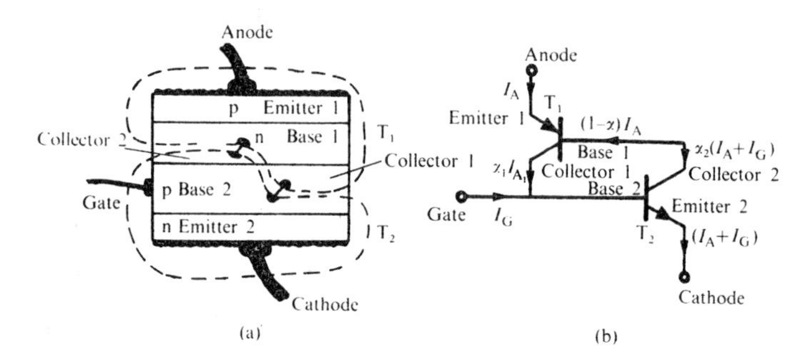

Fig.3.18. (a) A thyristor can be partitioned into two closely coupled transistors, one n-p-n and the second p-n-p: (b) the currents in the two complimentary transistors.

One way of regarding a thyristor is as two closely linked transistors in the same block of semiconductor (Fig.3.18(a)). The current between base 1 and collector 2 can be written in two ways, which must be equal (Fig.3.18(b))

$$(1 - \alpha_1)I_A = \alpha_2(I_A + I_G).$$

Hence

$$I_A = I_G\alpha_2/(1 - \alpha_1 - \alpha_2).$$

If $\alpha_1 + \alpha_2 = 1$, then I_A may have a finite value, even though I_G is zero. The value of α for a normal transistor usually exceeds 0·9, so the required values of α_1 and α_2 in the ON condition are low, and occur because the two component transistors run in the saturated mode (see page 79). In the OFF condition the short life of injected carriers will reduce ($\alpha_1 + \alpha_2$) to less than unity, so that the anode current is small until switched on by gate current.

Those parts of the thyristor nearest the gate connection turn ON first, and the high density electron-hole plasma necessary to fill the traps spreads by diffusion from the area where it was first formed. As a result there has to be a delay after a gate pulse before a thyristor can pass a high current. A maximum rate of rise of anode current is part of the device specification for a thyristor.

The switch-off time for a thyristor is governed by the need to wait for the plasma to decay to a low enough density for traps once more to be unoccupied. This may take tens of microseconds, and until it has occured, the voltage across the thyristor must be low, lest the thyristor switch ON again. As a result, thyristors are limited to relatively low operating frequencies, being excellent for controlling power at 50 Hz.

THE JUNCTION-GATE FIELD-EFFECT TRANSISTOR

The junction-gate field-effect transistor, or JUGFET, consists of a conducting channel with drain and source ohmic contacts at the two ends. The channel passes under p-n junction gates (Fig.3.19(a)) to which a control voltage can be applied relative to the source.

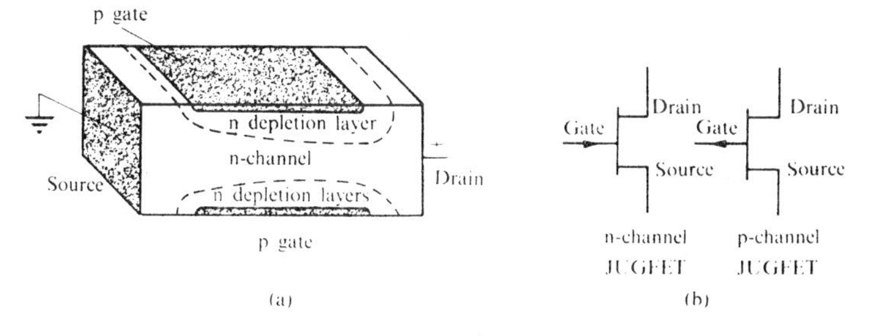

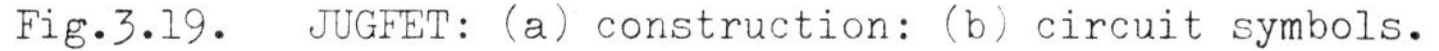
Fig.3.19. JUGFET: (a) construction: (b) circuit symbols.

Fig.3.8(d)(p.63) shows an experimental set of curves for an n-channel JUGFET relating the drain current I_D to the drain-source voltage V_{DS}, with the gate-source voltage V_{GS} as a parameter. As the gate is made more negative, less current flows, 5 V being enough to stop current almost completely. To increase I_D to its maximum value for a given V_{GS}, V_{DS} has to be several volts or more, a noticeably larger value than was required on the collector of a bipolar transistor. Notice also the way the family of curves fan out from the origin rather than breaking from a common envelope as the curves for the bipolar transistor seem to do.

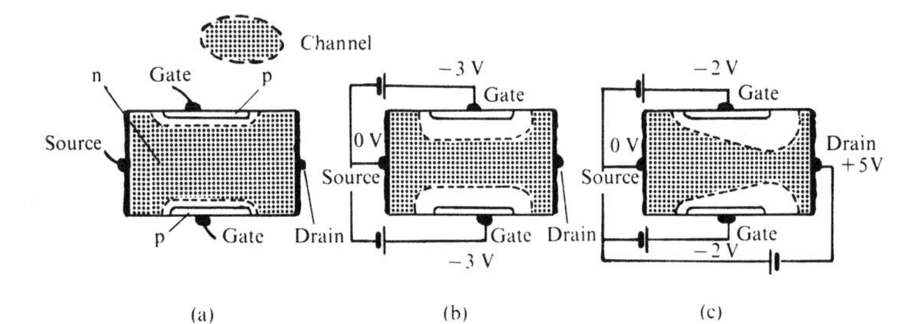

Fig. 3.20. n-channel JUGFET: (a) without bias: (b) with the gates biased negatively the depletion layers have expanded, reducing the channel width: (c) with negative gate bias and positive drain bias relative to the source, the channel becomes narrower near the drain.

The part of the semiconductor which can carry current is delimited by depletion layers whose boundary can be moved by varying V_{GS}(Fig.3.20(a),(b)). The device is thus a voltage-controlled resistor. The control voltage is applied across a reverse-biased p-n junction, so very little current flows, and a small power can control much larger powers.

Only two processes are needed to describe a JUGFET: the drift of majority carriers in an electric field, and the variation of the thickness of a depletion layer as the voltage across it is altered. The physics of the device is consequently simpler than the physics of the bipolar transistor, though this is compensated by the need to consider two dimensions in the simplest JUGFET analysis, while a one-dimensional analysis is a useful first stage in describing bipolar transistors.

To allow a reasonably simple analysis, we make two assumptions. A strict theory can show the assumptions are not well fulfilled, but the results fit experiment, and the ideas give a useful picture of the way the JUGFET works.

The two assumptions are 'the abrupt assumption' and 'the gradual assumption'. For the first we assume that the depletion layer ends abruptly at the boundary of the channel. In particular, in the depletion layer there are no free carriers to help carry the drain current, while in the channel the majority-carrier density is equal to the net doping density. Of course in a real JUGFET, the majority-carrier density would fall off smoothly at the edge of the depletion layer, but we shall assume the change is abrupt.

The gradual assumption refers to the rate at which conditions change in passing down the channel from source to drain. We assume that the channel and gate have a length L much greater than the width 2a of the JUGFET, so that the changes in the channel width are gradual, and only current flow parallel to the axis of the channel need be considered.

The channel potential varies continuously from source to drain, so the channel-to-gate potential difference V_y also varies, and consequently so does the undepleted channel width (Fig.3.20(c)). Take as an example an n-channel JUGFET, where the gates are p^+ and the channel is doped with N_d donors. If the width of the depletion layer is x, then at a distance y along the channel

$$x = (2\,\epsilon_r \epsilon_o\, V_y / eN_d)^{\frac{1}{2}}\ .$$

For simplicity, all potentials are internal potentials in the semiconductor. Externally measured potential differences will differ by a constant amount due to contact potentials.

The channel width is $2(a - x)$, and the electric field at a distance y along the channel from the source is $d(V_y)/dy$, so the current is

$$I = -2N_d e\mu_e \frac{d(V_y)}{dy} W(a - x) \, . \tag{3.40}$$

We can integrate eqn (3.40) along the length L of the channel:

$$\int_0^L I \, dy = -2 N_d e\mu_e W \int_0^L (a - x) \frac{dV_y}{dy} dy \, ,$$
$$I|y|_0^L = -2 N_d e\mu_e Wa \int_{V_{SG}}^{V_{DG}} (1 - \frac{x}{a}) dV_y \, . \tag{3.41}$$

If V_0 is the gate-channel voltage difference when $x = a$, then

$$V_0 = eN_d a^2/2\epsilon_r\epsilon_o$$

and (3.42)

$$x/a = (V_y/V_0)^{\frac{1}{2}}.$$

Carrying out the integral in eqn (3.41) gives

$$I = -\frac{2 N_d e\mu_e Wa}{L} \left\{ V_{DG} - V_{SG} + \frac{2}{3}\left(\frac{V_{SG}}{V_0^{\frac{1}{2}}}\right)^{\frac{3}{2}} - \frac{2}{3}\left(\frac{V_{DG}}{V_0^{\frac{1}{2}}}\right)^{\frac{3}{2}} \right\}$$

This can be written more compactly if we let

$$I_0 = -\frac{2}{3}\frac{N_d e\mu_e WaV_0}{L}$$

which is the value of I when $V_{SG} = 0$ and $V_{DG} = V_0$, and consequently is the maximum value that I can ever have.

$$\frac{I}{I_0} = 3\left\{\frac{V_{DG}}{V_0} - \frac{V_{SG}}{V_0}\right\} + 2\left\{\left(\frac{V_{SG}}{V_0}\right)^{\frac{3}{2}} - \left(\frac{V_{DG}}{V_0}\right)^{\frac{3}{2}}\right\} \, . \tag{3.43}$$

Eqn (3.43) holds while the depletion layers are narrower than the half-width of the JUGFET. The maximum potential difference between channel and gate will occur at the drain end of the channel, so when $V_{DG} > V_0$ we might expect the depletion layers to overlap and 'pinch-off' the channel. However Fig.3.8(d) shows that I_D saturates (i.e. stays constant) for large values of V_{DS} rather than falling, so we complete the theory by taking I_D for large V_{DS} to be the value reached when pinch-off occurs, that is when $V_{DG} = V_0$. Then for a pinched-off JUGFET, the characteristic equation is

$$\frac{I}{I_0} = 3\left\{1 - \frac{V_{SG}}{V_0}\right\} + 2\left\{\left(\frac{V_{SG}}{V_0}\right)^{\frac{3}{2}} - 1\right\} . \qquad (3.44)$$

As one would expect, when $V_{SG} = 0$, $I = I_0$. A value for V_0 can be estimated from the I_D versus V_{DS} curves. The current should reach a maximum when $(V_{DS} - V_{GS}) = V_0$, and for the example in Fig.3.8(d), this happens for V_{GS} between 0 V and -5·0 V, if we take V_0 to be 5·5 V.

Real JUGFETs have short gates, so the gradual assumption is not fulfilled, and numerical models show that the abrupt assumption is in error. However the predictions we make about the performance of the device are borne out usefully in practice.

JUGFET equivalent circuit

Above 'pinch-off' in our model the drain current depends only on V_{SG}, so the d.c. equivalent circuit is very simple (Fig.3.21(a)) requiring only a current generator between source and drain dependent on V_{SG} through a mutual conductance g_m.

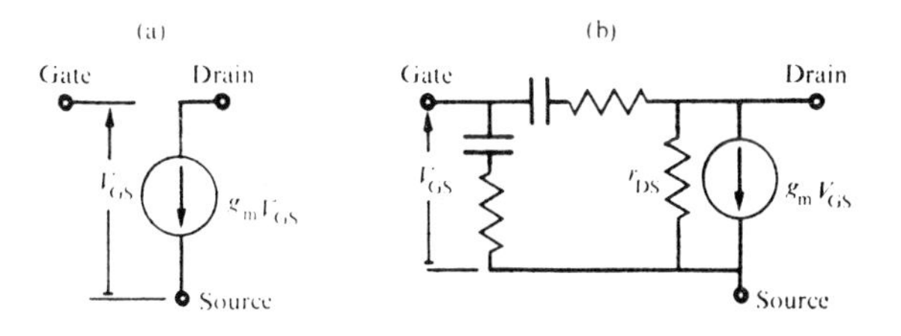

Fig.3.21. JUGFET equivalent circuits: (a) the simplest equivalent circuit for a JUGFET: (b) a more complete equivalent circuit.

To relate g_m to eqn (3.44), we must find (dI/dV_{SG}).

$$g_m = \frac{dI}{dV_{SG}} = \frac{-3I_0}{V_0}\left\{1 - \left(\frac{V_{SG}}{V_0}\right)^{\frac{1}{2}}\right\}. \qquad (3.45)$$

To make an equivalent circuit fit a device rather than our theory, the slight increase of I with V_{DS} may be represented by a resistance between drain and source. The d.c. current to the gate is probably too small to be worth including. The depletion layer capacitance of the gate-channel p-n junctions should be included, divided into two parts, one between gate and source, and the other between gate and drain. The currents through these capacitances must flow through the bulk semiconductor which has resistance, so a reasonably full equivalent circuit might be as in Fig.3.21(b).

An estimate of the highest frequency that a JUGFET can amplify may be obtained in several ways. The transit time τ_t of a carrier down the channel gives one measure. The simplest estimate of τ_t comes from assuming a constant field V_0/L along the device so that $\tau_t = L^2/\mu V_0$. When the expression for V_0 (eqn (3.42)) is inserted, we have

$$\tau_t = 2L^2\epsilon_r\epsilon_0/e\mu N_d a^2. \qquad (3.46)$$

The maximum frequency of operation f_{max} will be about $(1/2\pi\tau_t)$ so that

$$f_{max} = \frac{e\mu N_d a^2}{4\pi L^2\epsilon_r\epsilon_0} \qquad (3.47)$$

Another method of analysis is to treat the channel as a resistor through which current charges the gate-channel capacitance. Taking an average value of the channel width to be a in a device of width 2a, the channel resistance R is $L/e\mu N_d Wa$, and the gate to channel capacitance C is $2WL\epsilon_r\epsilon_0/0{\cdot}5a$.

The time constant, $CR = 4L^2\epsilon_r\epsilon_0/e\mu N_d a^2$, leads to an expression for f_{max}

$$f_{max} = e\mu N_d a^2/8\pi L^2 \epsilon_r \epsilon_o \qquad (3.48)$$

which has the same form as eqn (3.47), the difference by a factor of 2 being accounted for by the roughness of the approximations that have been made.

The factors which would lead to a high f_{max}, can be inferred from the inspection of eqn (3.47) or (3.48). The JUGFET should be as short as possible, and steady progress has been made in this direction. The mobility should be high, but only a new material can help here, and V_0 should be large, a price a user may not be prepared to pay.

The JUGFET as a variable resistor

The JUGFET characteristic (Fig.3.8(d)p.63) shows a fan of curves for small values of V_{DS}. The channel resistance is being controlled by the gate-source voltage, and the JUGFET can be used as an electronically variable resistance if drain-source voltages are well below pinch-off. An expression for the resistance r_D comes by finding (dI/dV_{DS}) for constant V_{GS}.

$$\frac{1}{r_D} = \frac{dI}{dV_{DS}} = \frac{\partial I}{\partial V_{SG}} \frac{\partial V_{SG}}{\partial V_{DS}} + \frac{\partial I}{\partial V_{DG}} \frac{\partial V_{DG}}{\partial V_{DS}} .$$

Since V_{GS} is constant $\partial V_{GD}/\partial V_{SD} = 1$, and $\partial I/\partial V_{GS} = 0$, so that

$$\frac{1}{r_D} = \frac{\partial I}{\partial V_{DG}} = \frac{3I_0}{V_0}\left\{1 - \left(\frac{V_{DG}}{V_0}\right)^{\frac{1}{2}}\right\} ,$$

which can be compared with the expression for g_m (eqn(3.45)) on the saturated part of the same curve. The coincidence can be used as a test of the theory, but is not of fundamental importance; other approximations would predict that the two quantities had similar but not identical values.

JUGFETs are preferred to bipolar transistors when a high input impedance or a high output voltage are required. On the other hand, bipolar transistors are currently faster, and more linear, so retain many small-signal applications.

LIGHT-EMITTING DIODES

Light-emitting diodes (LEDs) have become familiar as the 'answer in lights' on pocket calculators, and as the display in many other digital instruments. These LEDs and infrared-emitting diodes are p-n junction diodes whose construction and material are selected to produce light when current is passed through the diode in the forward direction. Electrically, the theory of pp58-59 applies without modification; for instance voltage-current relations will be of the form $I = I_r\{\exp(eV_{ext}/\kappa T)-1\}$, and in use very little current will flow until V_{ext} is more than some minimum value. For diodes emitting visible light, a forward bias of 1·0 V - 1·5 V is usually required. The intensity of the light output is found to be proportional to the current.

The basic process occurring is the emission of a photon when an injected electron (or hole) recombines, so in considering the physics of LEDs, attention is focused onto the possible recombination processes in semiconductors and the way in which the recombination energy is dissipated. Most recombinations occur within two diffusion lengths of the junction, so the light will emerge from this region. One possible transition is directly from the conduction band to the valence band (Fig.3.22(a)) when the energy at the peak of the spectrum will be about equal to the band-gap energy + κT. Other transitions may occur by stages, the energy of any photon emitted being less than the band-gap energy (Fig.3.22(b),(c),(d)). We can thus recognize two ways of controlling the energy of emitted photons. We can select a semiconductor with the right band gap, or we can add impurities to provide energy levels in the forbidden band. In practice both methods are employed.

A further problem is to arrange that a high enough fraction of the transitions go by a route where a photon is emitted, and that non-radiative transitions are few. The distinction between

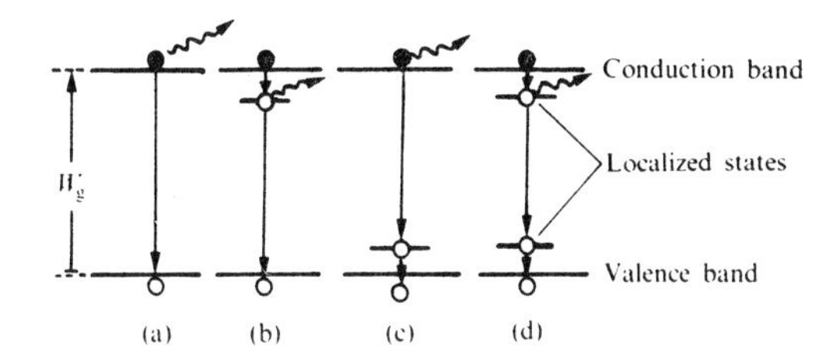

Fig.3.22. Transition processes for electrons. The photon is shown leaving the initial state of the radiative transition.

a direct-gap and an indirect-gap semiconductor is important here.

In a direct-gap semiconductor (e.g. GaAs, InSb) the minimum of the conduction band and the maximum of the valence band are both at the same value of k, so that transitions such as α in Fig.3.23(a) require no change of momentum, and all the released energy can be carried away by the low momentum photon. In an indirect-gap semiconductor (e.g. Si, Ge, GaP), the minimum of the conduction band is not at k = 0, so that a transition such as β in Fig.3.23(b) must involve some other particle to conserve momentum. The other body is usually a phonon, which also carries away the energy, so the transition in non-radiative. Transitions like γ are likely to be rare as there will be few electrons occupying the high energy k = 0 region of the band.

A photon is only useful when it has emerged from the diode. Absorption in the semiconductor, and total internal reflection at the surface reduce the effective output by a factor of up to one hundred in some devices.

Some examples from the many types of light-emitting diodes are as follows.

(a) GaAs infrared-emitting diodes. These emit at 1·4 eV in the near infrared by a band-to-band transition across the 1·4 eV direct band gap of GaAs. The internal quantum efficiency (fraction of injected electrons recombining with emission of

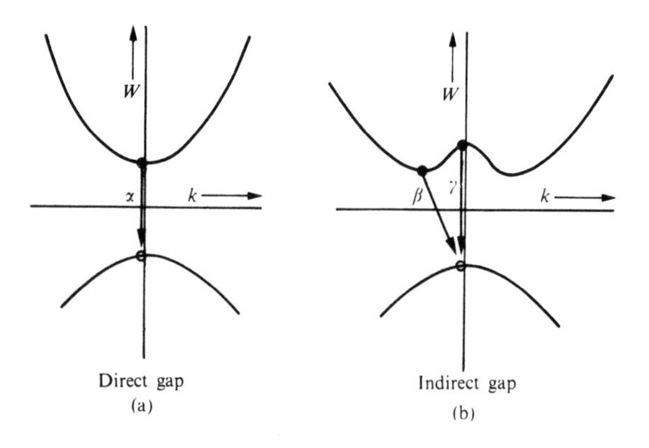

Fig.3.23. Electron transitions shown on an electron energy versus wave number diagram for semiconductors having a direct gap and an indirect gap.

a photon) may reach 15 per cent, but the external efficiency (infrared power out/electrical power in) is only about 1 per cent because the photons are readily reabsorbed.

(b) GaP doped with Zn and O. These emit red light of 1·7 eV energy. The band gap of GaP is 2·25 eV, and is indirect. The Zn and O, when on adjacent lattice sites, form an unusual neutral impurity energy level. The external efficiency is about 2 per cent because the chance of reabsorption is low.

(c) GaP doped with N. These emit a broad band of light whose appearance to the human eye changes from red to green as the amount of N is increased. The N may form a band of states in the forbidden energy gap which broadens as the N atoms come closer together. Efficiency is low, but as the eye is sensitive to yellow and green, the brightness is acceptable.

(d) $GaAs_x P_{1-x}$ doped with moderate amounts of N. As long as x is less than 0·5 the band structure is that of GaAs with a direct gap. The band gap increases as the fraction of P increases, and at the same time the colour of the light changes from red through to green.

LASERS

In this section we shall consider the physics of lasers, and then the semiconductor junction-diode lasers that have been developed from LEDs.

We start with the idea that the transition of an electron between two energy levels S_1 and S_2 can occur in three ways, ignoring the possibility of interactions with other particles or with quasi-particles such as phonons.

(a) Upwards, by absorption of a photon, at a rate given by

$$dN_2/dt = B_1 U(\nu) N_1 ,$$

where B_1 is a constant, N_1 and N_2 are the density of occupied S_1 and S_2 levels, and $U(\nu)$ is the density of radiation of frequency ν.

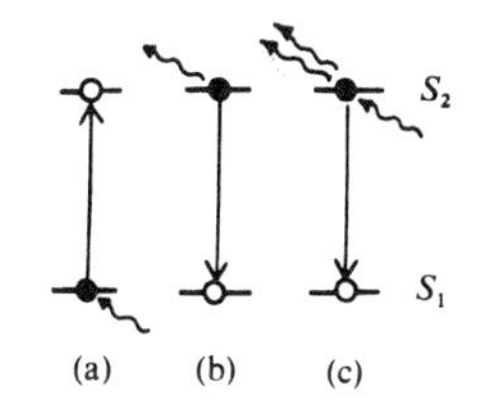

(b) Downwards, by spontaneous emission of a photon, at a rate

$$dN_2/dt = -AN_2 .$$

(c) Downwards, by stimulated emission of a photon. This happens when an excited state is kicked into emission by the field of a passing photon. The transition rate is

$$dN_2/dt = - B_2 U(\nu) N_2$$

The emitted photon is a replica of the photon which stimulated it.

The total rate of change of N_2 is

$$dN_2/dt = - AN_2 + B_1 U(\nu) N_1 - B_2 U(\nu) N_2 .$$

In a steady state $dN_2/dt = 0$, and hence

$$U(\nu) = \frac{A}{B_1(N_1/N_2) - B_2} .$$

One example of a steady state is thermodynamic equilibrium, where the Boltzmann relation $N_1/N_2 = \exp(-eV/kT)$ relates N_1 and N_2 to eV, the difference in energy between S_1 and S_2. Thus in thermodynamic equilibrium

$$U(\nu) = \frac{A}{B_1 \exp(eV/\kappa T) - B_2} .$$

However, this must be equivalent to the Planck formula for radiation density in any equilibrium situation at a temperature T

$$U(\nu)d\nu = \frac{8\pi\ h\nu^3}{c^3} \frac{1}{\exp(h\nu/\kappa T)-1}\ d\nu$$

If the two formulae are to match, $B_1 = B_2(=B)$ and $A/B = 8\pi h\nu^3/c^3$. Because $B_1 = B_2$ the probability of a transition that depends on the presence of a photon is the same in both directions, the actual ratio being determined by N_1/N_2. The ratio A/B is proportional to ν^3, and hence at low (microwave) frequencies spontaneous transitions are rare, while at high (visible) frequencies, stimulated transitions are relatively less frequent, and lasers more difficult to operate.

In a laser, two techniques are used to make the stimulated downward transition the dominant process.

(a) Population inversion, which means arranging for a high energy state to be more fully occupied than some state of lower energy. As a result $BU(\nu)N_2 > BU(\nu)N_1$, so that more photons are emitted by stimulated emission than are absorbed. To achieve population inversion requires a major shift away from equilibrium, and a high energy flow.

(b) Photon confinement, so that high photon densities can be reached. When $U(\nu)$ is high enough $BU(\nu)N_2 > AN_2$, and stimulated emission becomes more frequent than spontaneous emission.

A semiconductor junction-diode laser achieves population inversion by strong forward bias of a very heavily doped p-n junction, and a high photon density by mirrors deposited on two opposite polished faces of the semiconductor chip. The critical transition can be any of those indicated in Fig.3.22.

The requirement for population inversion in a semiconductor must take into account the spread of energies of the relevant states near the edges of the conduction and valence bands. As on pp , the rate of transition between the two bands is proportional to the number of candidates ready to make a transition, and to the number of places they may fill.

The rate of transition upward is

$$f_v A_v (1 - f_c) A_c \; U(\nu) B,$$

while the transition rate downward by stimulated emission is

$$f_c A_c (1 - f_v) A_v \; U(\nu) B,$$

where A_v and A_c are the effective densities of states in the valence and conduction bands, and f_v and f_c are the fractions of the states filled by electrons. We are assuming that $U(\nu)$ is high enough to allow spontaneous transitions to be ignored. No exponential factor distinguishes upward and downward transition rates as the same photon density drives both. If stimulated emission is to be more frequent than absorption then

$$f_v A_v (1 - f_c) A_c < f_c A_c (1 - f_v) A_v$$

or

$$f_c > f_v \quad .$$

Thus for the laser to operate, there have to be more electrons near the bottom of the conduction band than near the top of the valence band. This can be achieved by doping the junction so

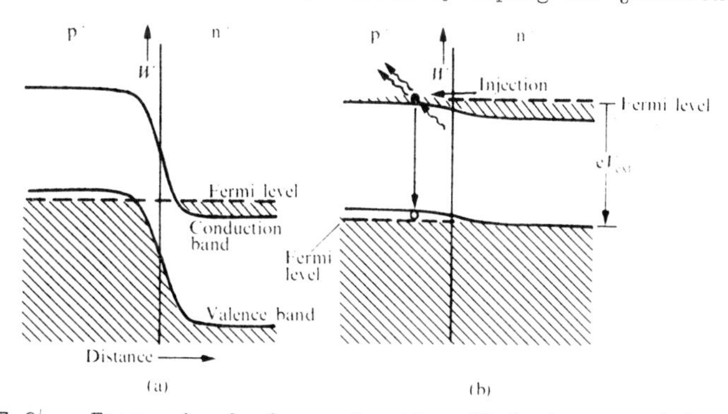

Fig.3.24. Energy bands for a junction-diode laser: (a) without bias: (b) with enough bias to invert the electron population in the p^+ region near the junction.

heavily that the Fermi level is above the bottom of the conduction band in the n-side, and below the top of the conduction band in the p-side (Fig.3.24(a)), and applying a forward bias about the size of the barrier potential which will be a little more than the band gap, as in Fig.3.24(b), which represents a GaAs laser with an infrared output. The current density of such a laser at 300 K is 6×10^8 A m^{-2}; a chip 1 mm square would require 600 A.

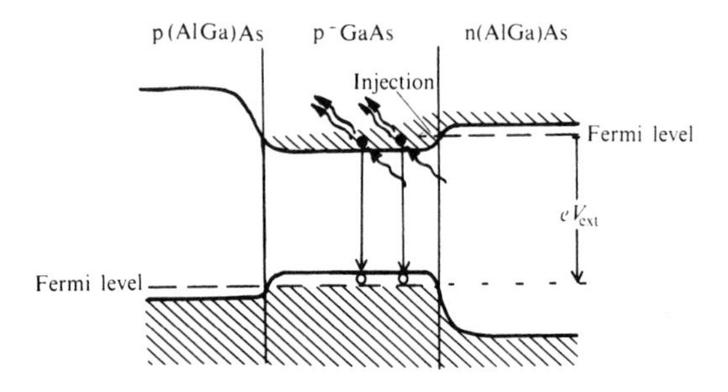

Fig.3.25. A double-heterojunction laser with working bias. Injected electrons are trapped in the p^+ GaAs conduction band.

By using a three-layer structure, where only the meat in the sandwich is GaAs, the threshold current density is reduced by a factor of 30 (see Fig.3.25). The outer two layers are $Al_xGa_{1-x}As$ which has a band gap of 2 eV by comparison with 1·4 eV for GaAs. The p-$Al_xGa_{1-x}As$ confines the injected electrons to the p-GaAs, so that an adequate density can be built up with a lower current. A second effect comes from the lower refractive index of the two $Al_xGa_{1-x}As$ layers, which help to confine the photons to the central active region. The reduction in the current required to drive the laser makes continuous room temperature operation possible, and greatly extends the life of the lasers, which tends to be short as a result of defects introduced by thermal stresses.

IMPATT DIODES

It has been found that many kinds of diodes which embody a p-n

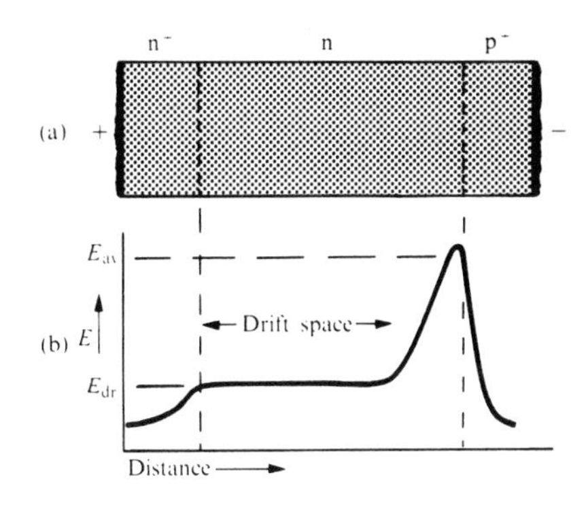

Fig.3.26. IMPATT diode: (a) schematic structure: (b) variation of average electric field E along the diode.

junction can act as microwave oscillators when placed in a resonant circuit and reverse-biased just beyond the edge of avalanche breakdown. The name 'IMPATT diode' derives from a fuller title IMPact Avalanche Transit Time diode.

One of these diodes is sketched in Fig.3.26(a). The total field takes into account both the barrier potentials and the applied voltage (Fig.3.26(b)). In the central n-region, the field is somewhat above the value E_{dr} required to make the electrons drift at their maximum speed, while at the n-p^+ junction, the field rises above the value E_{av} required for an electron-hole avalanche to start. The device is in a resonant circuit, so there will be a repeating cycle of events.

After the peak field rises above E_{av} the avalanche builds up, reaching its maximum only when E has passed its own peak and fallen again below E_{av}. The holes produced in the avalanche rapidly reach the p^+-contact, taking no further part in the process, but the electrons are released into the n-region where they are not neutralized by either donors or holes. As a result the field must rise where the injected electrons are present, and there is less voltage difference to develop a high field across the n-p^+ junction, so that the avalanche ceases. The electrons drift at their maximum velocity across the n-region, and current continues to flow in the external circuit while they do so. Thus there are two delays between the voltage and the current waveforms, and if the frequency of the resonant circuit is suitable, the current can be between $\frac{1}{2}\pi$ and $\frac{3}{2}\pi$ out of phase with the voltage. It is therefore feeding a.c. power to the circuit rather than absorbing it as a normal resistive load does.

To make a high frequency, the drift length and avalanche delay must be short. Power output of 2 W at 7 GHz with a d.c. power supply of 100 V and an efficiency of 6·5 per cent is available commercially, while 200 μW at 361 GHz has been achieved in experiments.

Other modes of oscillation are seen in practice, some of which are of much higher efficiency and hence of interest to potential users.

PROBLEMS

3.1. An abrupt Si p-n junction is doped with $10^{21}\,m^{-3}$ B atoms in the p-region and $10^{20}\,m^{-3}$ P atoms in the n-region. Calculate, at 300 K:

(a) V_b, the barrier potential;

(b) the depletion-layer thickness in each side when -10 V is applied;

(c) the capacitance with - 10 V bias, if the area is $10^{-8}\,m^2$;

(d) the minimum voltage for avalanche to start.

(For part (d), assume the avalanche occurs when the field exceeds $3{\cdot}0 \times 10^7$ V m^{-1} at some point.)

3.2. A reverse-biased p-n junction diode may be used as a voltage-controlled capacitor in a tuned circuit where the resonant frequency f is proportional to (diode capacitance)$^{\frac{1}{2}}$.

If a symmetrical Si step junction diode has a net doping density on each side of the junction of 10^{22} m^{-3}, by what factor does f vary as the bias is changed from -1 V to - 10 V?

3.3. A Si step junction is initially unbiased. A current of 1 mA is passed through it, so as to make it steadily more negatively biased. How long does it take for the bias voltage to reach -10 V? Take the doping density on each side of the junction to be $10^{21} m^{-3}$ and the area of the junction to be $10^{-6} m^2$. (Hint: find the charge in the diode at -10 V bias.)

3.4. Design a diode which will pass 10 mA when a forward bias of 0·5 V is applied. (Note: diffusion distances can be controlled in manufacture - a value in the range 10^{-5} - 10^{-4} m would be reasonable.)

3.5. For the Si and Ge transistors whose characteristics are shown in Fig.3.8 (p 63) take a quiescent working point at V_{CE} = 5 V and I_C = 25 mA and estimate the hybrid-π parameters g_m, r_π, and r_0, and also β.

3.6. Using the hybrid-π model of a bipolar junction transistor with r_x, r_μ and C_μ neglected, find the relation between τ_t and the frequency when the current gain is 3 dB down on its d.c. value for a common emitter amplifier fed from a voltage source.

3.7. If two transistors differ only in that one has a base width which is 90 per cent of the base width of the other, what will be the difference between the base-emitter voltages needed to maintain the same current flowing in the two transistors, under reasonable forward bias?

4. Surfaces and interfaces

One kind of interface - that between p- and n-type semiconductor- has already been analysed. Other surfaces also have importance in electronic devices.

In this chapter there is first a treatment of surfaces on the atomic scale, then a discussion of charged layers at an insulator-semiconductor junction, leading to an analysis of the metal-oxide-semiconductor (MOS) transistor. Metal-metal junctions introduce metal-semiconductor junctions, used in Schottky-barrier devices. Junctions between different semiconductors are now becoming important, and are briefly discussed.

SURFACE PHYSICS

The surface of a crystal exposed by breaking it in half is clearly not the same as the inside of the crystal. Several methods of analysis confirm this.

The bond model of a crystal describes a surface as having one unjoined bond per atom, or about 10^{19} bonds per square metre (see Fig.4.1). The bonds can react with neutral particles, atoms or molecules or with charged particles - particularly electrons.

A wavefunction designed to fit an infinite periodic lattice clearly is incorrect outside the crystal. Wavefunctions have been specially devised to fit a periodic potential that stops abruptly. The energy levels of these wavefunctions (Tamm states) differ from the bands of allowed levels which fitted the inside of the crystal. They allow electrons to exist near the surface with energies between the valence and conduction bands. Thus electrons bound on unjoined bonds have a quantum-mechanical description as occupied surface states.

The presence of charges in surface states changes the electrical potential and hence changes the level of the bands of

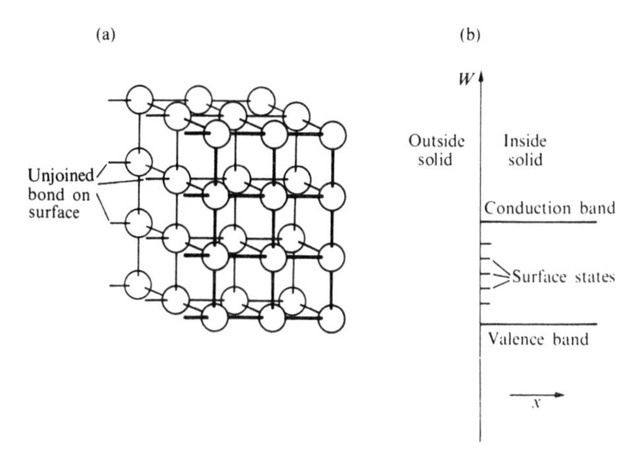

Fig.4.1. Surface of a crystal: (a) the bond model: (b) the energy-band model.

allowed energy, so that the band model of the crystal also reflects the special surface effects. The bands are bent, and much of this chapter is about band-bending. Two points may be noted, (a) half a p-n juntion is often worth comparing with a surface, and (b) if bands are bent, then $d^2V/dx^2 \neq 0$, so that by Poisson's equation the region is not neutral.

A casually prepared semiconductor surface can have a surface-state density approaching the value of 10^{19} states per square metre quoted earlier, and in addition may have an unknown, unstable assortment of chemicals on the surface. Modern semiconductor technology can prepare surfaces which have far fewer surface states, far lower surface charge density in those states, and a stable chemical nature.

The stages in preparing a surface might go as follows: cut the grown crystal with a diamond saw, polish the surface to a mirror finish; etch off a thin surface layer to remove strained material containing large numbers of structural defects; and finally, after the required doping impurities have been diffused in, deposit a stable insulating layer on the semiconductor surface.

In much of the following discussion we shall ignore the surface states completely, thus simplifying the analysis enormously while keeping it accurate enough to be useful. A few devices use a specially prepared active surface, and for these we make a detailed examination of the surface, but in general we concentrate on the regions near the surface, while ignoring effects peculiar to the last few layers of atoms. Surfaces are of interest in many fields - catalysis, friction, and detergents are examples - and a full study of surfaces would draw on contributions from many sciences.

WORK FUNCTION, FERMI LEVEL, AND ELECTRON AFFINITY

The work function ϕ of a crystal surface is the amount of energy required to remove an electron from the middle of the crystal to infinity. There is a further step required to make this definition precise - we must say which electron. For a metal, the choice is easy: we take an electron at the top of the filled levels (Fig.4.2). At room temperature the levels change from being nearly full to nearly empty in about 100 meV. A typical work function is 4 eV, so for many purposes the Fermi function is a step function, and we can define the work function as the energy required to remove an electron from the Fermi level inside the crystal to infinity, or to give the same idea another name, to the vacuum level.

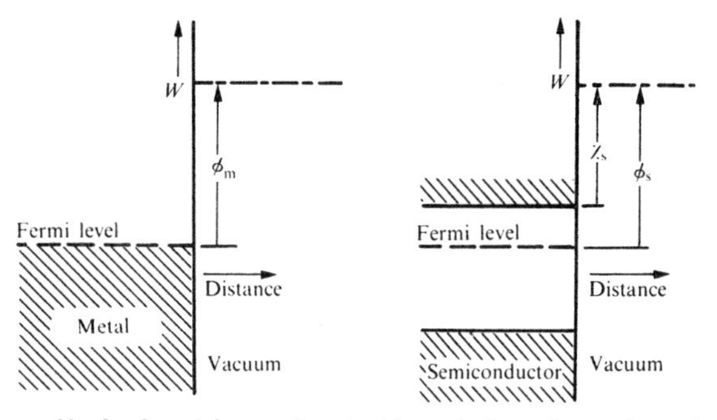

Fig.4.2. Work functions for both metal and semiconductor are measured from the Fermi level.

In a semiconductor there probably are no states at or near the Fermi level, so an electron cannot be removed from them. Nevertheless, we measure the work function from the Fermi level for a semiconductor, just as we did for a metal. To find how much energy an electron near one of the band edges has to be given to emerge from the crystal we must correct the work function by the difference between the Fermi level and the relevant band edge. It is found in practice that the work function of semiconductors varies as the doping is varied, n-type Si having a lower work function than p-type Si. This variation is nearly all accounted for by the change in the relative positions of the Fermi level and the band edges. The energy required to allow an electron to escape from the bottom of the conduction band is the same for p-type and n-type material. A constant property is useful to work with - we call this property 'electron affinity' χ, (Fig.4.2).

For a metal	Electron affinity = work function. Both are measured from the Fermi level.
For a semiconductor	Electron affinity is measured from the bottom of the conduction band to the vacuum level and is a constant for a given crystal. Work function is measured from the Fermi level to the vacuum level and depends on the doping and on the kind of crystal.

BAND-BENDING AND FIELD-INDUCED SURFACE LAYERS

On the positive plate of a charged capacitor there is a positive surface-charge density and on the negative plate an equal negative charge (Fig.4.3). By considering what states are available for these charges to occupy we

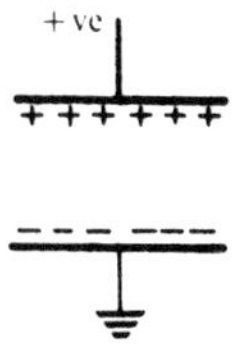

+ve

Fig.4.3

can establish how thick the surface layer is which they cause. In a metal there are plenty of states into and from which the extra electrons can be transferred, and the field falls from a high value outside the metal towards zero in a few atomic layers. In a semiconductor there are only a small number of states which, at acceptable energies, can be filled or emptied, and if a large charge has to be accommodated, then the charge on the capacitor extends much further into the crystal.

In Fig.4.4(b), p-type semiconductor forms the negative part of a capacitor. The usual electrostatic convention that surface charges exist in negligibly thin layers is on this occasion being refined by a more detailed analysis. The charge density at the geometrical surface is zero, so that the electric displacement $D = \epsilon E$, where $\epsilon = \epsilon_r \epsilon_o$, is continuous in the mathematical sense on entering the surface. As ϵ is different inside the semiconductor, the electric field inside is not the same as the electric field outside. The displacement falls to zero deep in the semiconductor from its value D outside. Hence the total charge density per unit area, S in the whole surface layer is, by Poisson's equation, equal to D. The bands bend in this case to lower energies near the surface. This removes the valence band from the Fermi level so that there are now holes in the valence band near the surface, though the acceptors stay ionized. The volume-charge density is that of the acceptors, and a useful estimate of the thickness t of this surface layer, if the acceptor concentration is N_A is

$$teN_A = S = \epsilon E .$$

This sort of surface layer is called a ***depletion*** layer, and is similar in many respects to the depletion layer on one side of a p-n junction. For instance, the relation between the voltage V across the depletion layer and the thickness is

$$V = \frac{t^2 e N_A}{2\epsilon_r \epsilon_o}$$

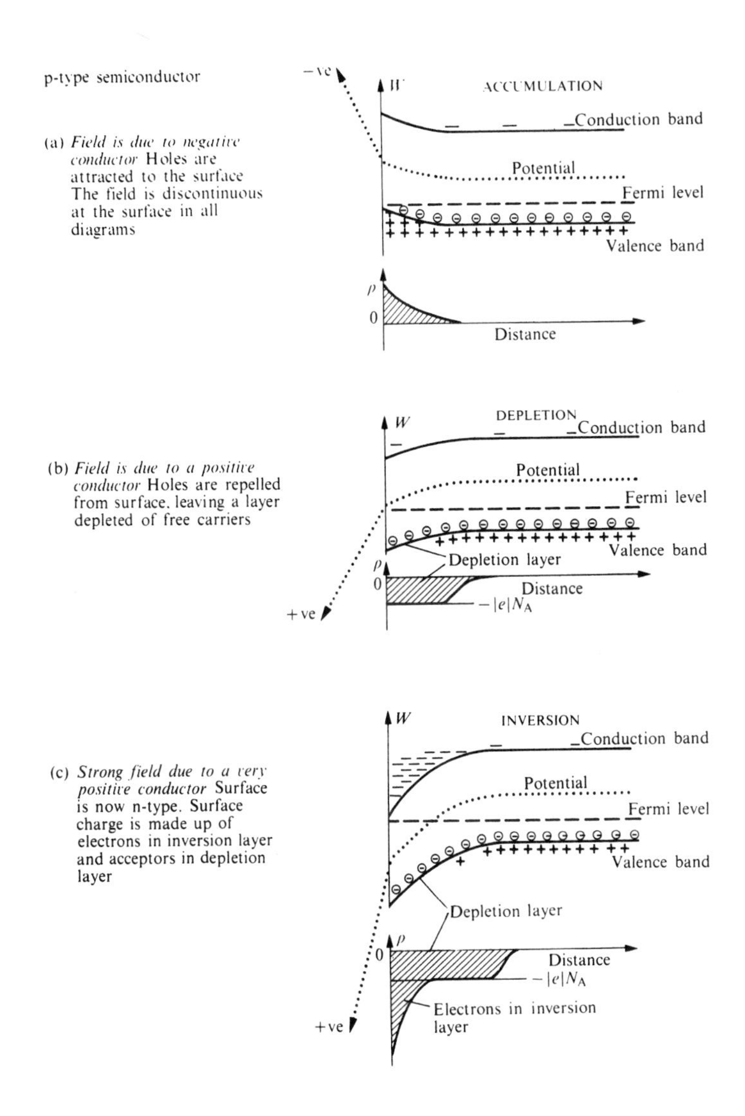

Fig.4.4. The three types of surface layer for a p-type semiconductor.

If p-type semiconductor is made the positive plate of a capacitor and hence the electric field is directed out of the surface, the bands are bent to more negative energies near the surface (Fig.4.4(a)) and there is an ***accumulation*** of holes near the surface. The charge density increases near the surface as the valence band approaches the Fermi level, so that the formula describing the thickness of the layer has to include the integral of a charge density which varies with position. The formula is thus more complex than that for the depletion layer.

A strong field into the surface of p-type semiconductor (Fig.4.4(c)), can cause the conduction band to become the nearer of the two bands to the Fermi level. The surface is then n-type, even though the bulk is p-type, and electrons occupy states in the conduction band near the surface in an ***inversion*** layer. Such a layer forms the conducting channel in many practical devices.

If instead of p-type semiconductor, n-type is used, corresponding effects occur for the opposite electric field direction. Thus, a field into the surface produces enhancement, a weak field out of the surface produces depletion, and a strong field out of the surface produces inversion. In the inversion layer the majority charge carriers would be holes.

Fig.4.4. needs understanding thoroughly. Check that the sign of the space charge fits with the curvature of the potential, and that charges agree with the gap between Fermi level and band edges. Drawing the corresponding diagrams for n-type semiconductors is a good exercise.

MOSFETs

The idea of using a semiconductor as one plate of a capacitor has been developed into the MOSFET (Metal-Oxide-Semiconductor Field-Effect Transistor) or MOS for short. IGFET (Insulated-Gate FET) is another name, and there are variations if the insulator is not an oxide.

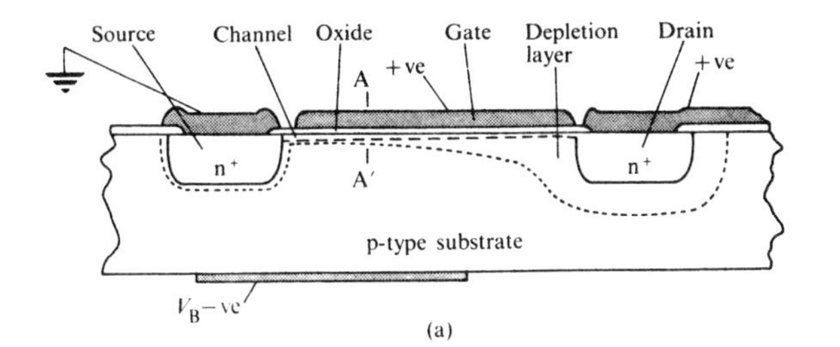

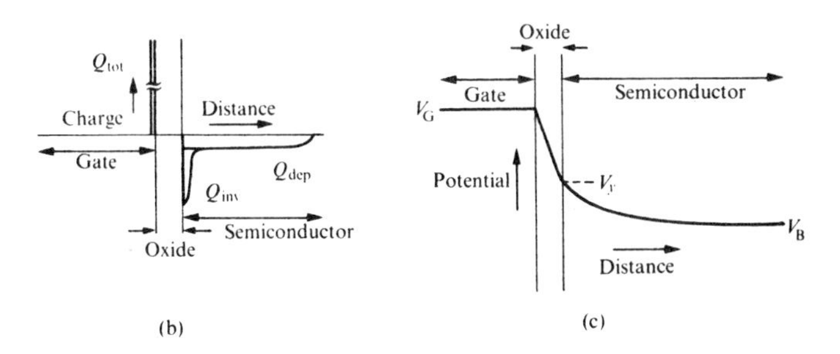

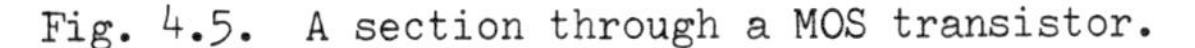
Fig. 4.5. A section through a MOS transistor.

One useful structure is shown in Fig.4.5. Substantial current only flows between the n^+ drain and source contacts when there is an n-type layer joining them, so that the surface has to be an inversion layer. The gate bias must therefore be positive relative to the source. If the gate bias has not formed an inversion layer, then one of the p-n^+ junctions will be reverse-biased, and only a small drain current can flow. The p-type substrate is lightly doped, but is not an insulator and is all at the same potential except in depletion layers, where it abuts onto the three n-type regions. It is biased negatively to both source and drain so that large currents do not flow.

In the normal set of bias conditions, the depletion layer between source and substrate is thin, the depletion layer between drain and substrate is thick, and that between channel and substrate makes a smooth transition between the two end conditions. The drain must be positive relative to the substrate, and hence is positive relative to the source, or at least only slightly negative.

By picking a special model for a MOS we can describe it with simple mathematics, though some features of the real device will not be accurately portrayed. The model and mathematics are closely related to those used in discussing the JUGFET (see pp 87-89). The gate and channel are long, and trapped charge and contact potentials are ignored. The method is to find the mobile charge and hence the conductivity of an element of the channel, and then find the voltage between source and drain by integrating the resistive voltage changes along the channel.

The fundamental statement describing the device (Fig.4.5(b)) is

$$Q_{tot} = Q_{inv} + Q_{dep},$$

where Q_{tot} is the total charge per unit area induced in an element of the channel, Q_{inv} is the charge per unit area in the inversion layer, and Q_{dep} is the fixed charge per unit area in the depletion layer. We need Q_{inv}, and find it by obtaining expressions for Q_{tot} and Q_{dep}.

$$Q_{tot} = \epsilon_{ox} E_{ox} = \epsilon_{ox}(V_G - V_y)/d \ ,$$

where ϵ_{ox} is the permittivity of the oxide, E_{ox} is the electric field in the oxide, d is the oxide thickness, V_G is the potential of the gate and V_y the potential of a point in the channel distance y from the source as in Fig.4.5(c). If V_B is the potential of the substrate, then from the analysis of a depletion layer with a voltage $(V_y - V_B)$ across it in a semiconductor of permittivity ϵ_s and acceptor density N_A,

$$Q_{dep} = \{-2\epsilon_s N_A\, e(V_y - V_B)\}^{\frac{1}{2}} \ .$$

Hence

$$Q_{inv} = \epsilon_{ox}\frac{(V_g - V_y)}{d}\{-2\epsilon_s N_A\ e(V_y - V_B)\}^{\frac{1}{2}}.$$

The current I is $\mu Q_{inv}(dV_y/dy)$, where μ is the mobility of the carriers in the channel.

As with the JUGFET, we integrate over the channel length, and find that

$$I_D = \frac{W\mu}{L}\left[(V_{DS}V_{GS} - \tfrac{1}{2}V_{DS}^2)\epsilon_{ox}/d - \tfrac{2}{3}(-2N_A e\epsilon_s)^{\frac{1}{2}}\{(V_{DS} - V_B)^{\frac{3}{2}} - V_B^{\frac{3}{2}}\}\right].$$

There are several points to note:

(a) The drain current I_D depends on three voltages V_{DS}, V_{GS}, and V_{BS}. The last of these is commonly constant, so the current can be presented graphically as a function of V_{DS} with V_{GS} as a parameter, as in Fig.3.8(c)(p. 63).

(b) The equation for I_D holds only as long as $Q_{inv} > 0$, this being the equivalent of the JUGFET not being pinched-off. Above this level I_D is roughly constant, though our theory fails to predict this.

(c) The mobility of the carriers near the surface may be lower than their mobility in the bulk material, as collisions with the surface become more frequent.

(d) The effects of the built-in voltages have been ignored, but can be included without difficulty.

(e) However, a short channel or a field-dependent mobility are much less easy to describe, and are likely to occur in practice.

The theory which has been outlined allows us to understand the interrelations between the various physical quantities, but a more exact method, probably using a computer, must be used to predict experimental curves like those in Fig.3.8(c)(p.63).

The characteristics in Fig.3.8(c) look similar to those for a JUGFET, and differ from those for a bipolar transistor. Notice that the parameter changing from curve to curve is the gate-source voltage not a current. The lines do not become horizontal in the

0 - 10 V range displayed; MOS devices can have V_{DS} up to 30 V.

If characteristic curves for other MOS devices are inspected, it will be seen that they are similar to Fig.3.8(c), but that some n-channel devices require zero gate bias to reduce the channel conductivity to zero, some require a few volts in the positive direction and some (like that in Fig.3.8(c) require a few volts negative. These differences are due to charge trapped in the insulator and at the insulator-semiconductor interface. The trapped charge is balanced by an opposite mobile charge in the semiconductor surface layer which may therefore be an accumulation, depletion, or inversion layer before any bias is applied to the gate, depending on the quantity and sign of the trapped charge.

From the point of view of a user the important distinction is between a device which passes current only when a gate bias is applied to enhance the channel conductance - an enhancement MOS, and a device which can pass current when the V_{GS} is zero, and which requires bias to stop current flow - a depletion MOS. Care must be taken to distinguish the two uses of the term 'depletion', in describing a surface layer and a device. Much effort on the part of manufacturers has gone into making devices where the trapped charge is constant in time and uniform from device to device. One line of research in this direction has been the use of insulators other than SiO_2.

A simple small-signal equivalent circuit for a MOS is shown in Fig.4.6, in the same orientation as Fig.4.5(a), so as to show which electric process each of the circuit elements represent. The circuit when in use is usually twisted round to look like Fig.4.7, but the circuit itself has not been modified. The capacitances represent the total capacitance from each terminal to each of the others. The resistance r_{DS} represents the change of I_D, the drain current passing along the channel, when V_{DS} is altered, and the current generator with mutual

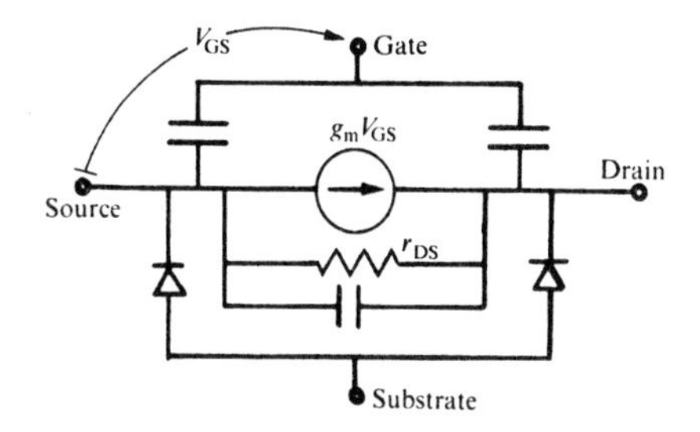

Fig.4.6. Equivalent circuit for a MOS transistor, drawn with the same orientation as Fig.4.5. The physical processes can be readily identified.

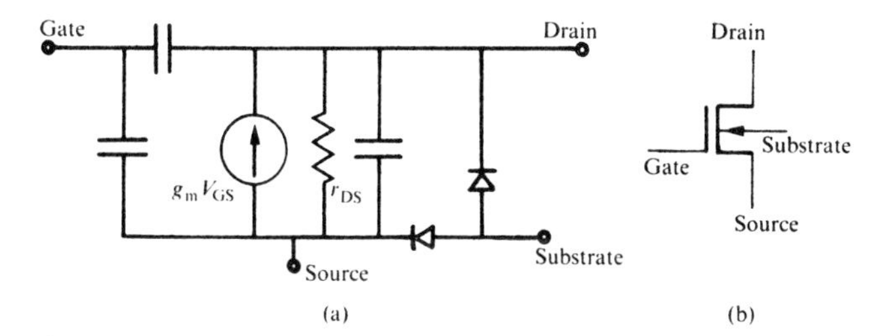

Fig.4.7. Equivalent circuit for a MOS transistor. This orientation is convenient for circuit analysis.

conductance g_m represents the change of I_D produced by a change of V_{GS}.

As an exercise, the reader could calculate values for r_{DS} and g_m for the MOS in Fig.3.8(c) for a bias point $V_{DS} = 5$ V, $V_{GS} = 0$ V.

In this simple equivalent circuit, there are no resistive paths from the gate to the rest of the circuit. This makes the point that the actual resistance is often so high that its admittance can be neglected. In some applications the high-resistance properties are being exploited to the full, and the designer or user might want to know just how high the resistance is. The equivalent circuit could be extended to show this; a value of 10^{13} Ω is possible as a gate resistance.

The MOS, as a more recent device than the bipolar junction transistor, has displaced the latter from some applications. Its high input resistance has been mentioned and explains why it is used where very small currents are to be measured. The MOS is used extensively in digital switching circuits. Integrate circuits can readily include MOS devices as their active components, and the large number of logical elements required for a computer may thus be included in a small space. However, for amplifying small signals, linear circuits are easier to build using bipolar transistors, which are thus preferred for some uses.

METAL-METAL JUNCTIONS

The effects occurring when dissimilar metals are brought together are worth studying because they help to explain the metal-semiconductor junction (which is examined in the next section) and also because they are of interest in their own righ

When the surfaces of two metals of different work functions ϕ_1 and ϕ_2 form the plates of a capacitor there will be an electric field between them. If the metals are in contact elsewhere, the Fermi levels will be at the same energy, and the voltage difference in the free space between the two metals will be $(\phi_1 - \phi_2)/e$. This potential difference can have observable effects. It can accelerate electrons, either a deliberately injected electron beam or photoelectrons released from one of the surfaces; and it causes there to be a charge on each capacitor plate. The Kelvin method of measuring work functions (differences of work function in fact) uses this effect (Fig.4.8 The distance between two plates is varied, and hence the field for a given potential difference. An ammeter in the circuit connecting the two plates will indicate a changing charge on the plates, except when the field is zero, which will occur when an external voltage in the circuit exactly cancels the potential difference $(\phi_1 - \phi_2)/e$. A measurement of the external voltage

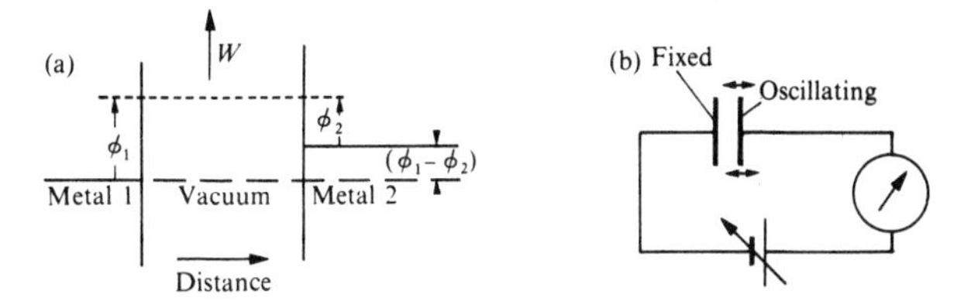

Fig.4.8. The Kelvin method for measuring work function differences: (a) when the field between the plates is zero, the energy difference between the two Fermi levels is $(\phi_1 - \phi_2)$; (b) the charge on the capacitor is independent of the distance between the plates only when the field is zero.

thus measures $\phi_1 - \phi_2$.

METAL-SEMICONDUCTOR JUNCTIONS OR SCHOTTKY BARRIERS

The effects that occur at the interface between a metal and a semiconductor depend on the work functions of the two materials, assuming that surface states do not play a dominant part. To understand this we examine a metal and semiconductor being brought into contact. Fig.4.9(a) shows the two materials well separated. Their Fermi levels are at the same energy; there is a small field in the space between them equal to $(\phi_m - \phi_c)/ed$.

When the metal and semiconductor are brought close enough, the potential difference cannot be taken up in the small space but produces a field in the semiconductor. We may say that the bands are bent (see p.107). When the materials touch, the whole of the work function difference $\Delta\phi$ is taken up by the band-bending (Fig.4.9(c)).

Another way to think about the contact is to say that the electron, in order to leave the metal, has to be given ϕ_m energy, of which ϕ_c is returned on entering the semiconductor. The difference $\Delta\phi$ appears as a change in potential in going from metal to semiconductor. The recipe for drawing diagrams such as Fig.4.9(c) is therefore as follows.

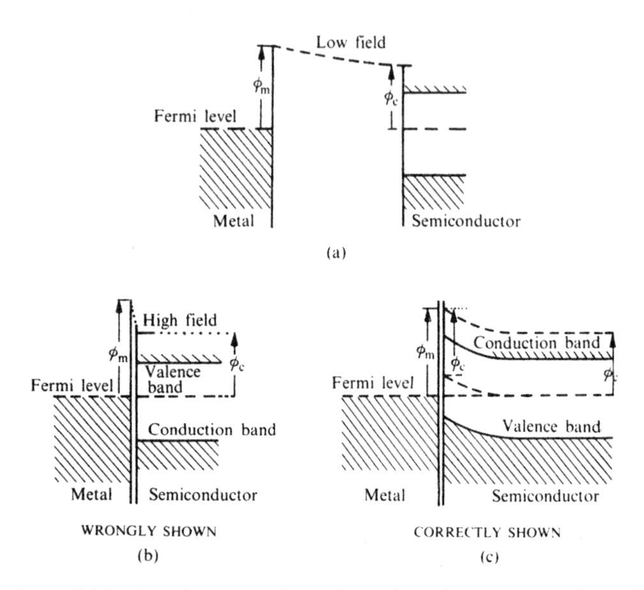

Fig.4.9. Effects when metal and semiconductor are brought together: (a) when metal and semiconductor are far apart the field between them is low; (b) the sharp change of field shown at the edge of the semiconductor is forbidden by Poisson's equation; (c) the difference of work function is taken up by band-bending in the semiconductor.

Start with the metal, draw the Fermi level for the semiconductor in the right place allowing for any external voltage, then draw the energy bands for the bulk semiconductors to fit round the Fermi level. Next locate the bottom of the conduction band at the surface by measuring up by ϕ_m from the metal Fermi level, measuring down by ϕ_c and correcting for the gap between conduction band and Fermi level in the semiconductor. This last stage can be more tidily expressed by saying that, at the surface, the bottom of the conduction band is $\phi_m - \chi_c$ above the metal Fermi level, where χ_c is the electron affinity of the semiconductor. Finally complete the bent bands in the semiconductor. Their thickness is once again governed by charges in the depletion layer through Poisson's equation. Notice that since the potential in the semiconductor is now no longer uniform, the carrier density will also vary. The local difference between the band edges and the Fermi level must be used to find hole and electron densities.

There are four cases of metal-semiconductor contact worth distinguishing, depending on whether the metal or semiconductor has the larger work function and whether the semiconductor is p-type or n-type. Two of these combinations lead to rectifying junctions and two to non-rectifying or 'ohmic' contacts. We shall examine one example of each.

Case 1. $\phi_m > \phi_c$, p-type semiconductor

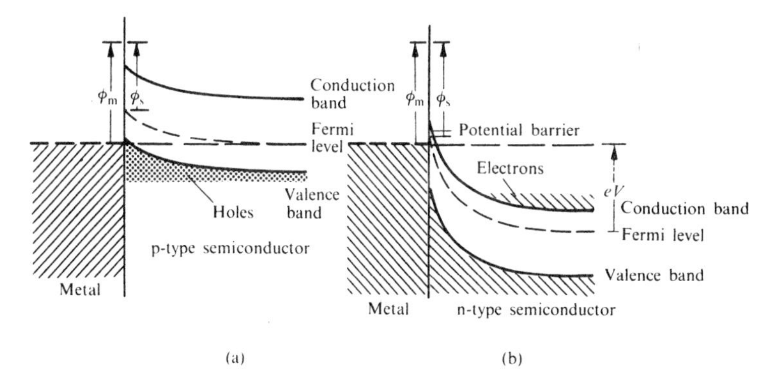

Fig.4.10. Examples of Schottky or metal-semiconductor interfaces: (a) $\phi_m > \phi_s$, p-type semiconductor, no external bias; hole flow is easy, the contact is 'ohmic'; (b) $\phi_m > \phi_s$, n-type semiconductor, diode reverse-biased. The electron flow is impeded by the potential barrier, and the contact is rectifying.

Fig.4.10(a) shows the energy-band diagram for zero bias. The interfacial layer is an enhancement layer, and the holes in it can exchange places readily with electrons in the metal. For small biases of either sign the hole flux will not be balanced and current will flow easily. The contact is known as an ohmic contact, and is one way of making good electrical connections to p-type semiconductor. An ohmic contact to n-type semiconductor can be made when the metal has the lower work function. (Surface states also are important in practical ohmic contacts.)

Case 2. $\phi_m > \phi_c$, n-type semiconductor.

When the contact is made between n-type semiconductor and a metal of higher work function, a potential barrier is set up between the two layers. The majority carriers in the semiconductor (electrons) now have to flow over a potential barrier of height $(\phi_m - \phi_c)/e$, and their flux is reduced by a factor $\exp\{(\phi_m - \phi_c)/\kappa T\}$ by comparison with a case where there is no barrier. When an external potential V is applied the barrier is $\Delta\phi + eV$, so the exponential is altered to $\exp(\Delta\phi/\kappa T).\exp(eV/\kappa T)$. The situation has similarities to a biased p-n junction, but although a V-I relation of the form

$$I = I_o\{\exp(eV/\kappa T)-1\}$$

seems likely, there are difficulties in establishing a value for I_o.

Metal-semiconductor diodes made in this way are known as Schottky-barrier diodes. Contact between p-type semiconductor and metal of a lower work function also produces a Schottky-barrier diode.

The injection of carriers occurs in the same way with p-n junction diodes and Schottky-barrier diodes, but the recombination of the injected carriers can occur more quickly in the metal than in the semiconductor. The Schottky - barrier diode is consequently very fast, and many microwave detector or mixer diodes are Schottky-barrier diodes. Point-contact diodes, where a metal point is welded to a semiconductor, have long been used for high-frequency rectification, and may have been Schottky barrier diodes, though without the fact being realized.

SEMICONDUCTOR-SEMICONDUCTOR CONTACTS OR HETEROJUNCTIONS

It is possible to make a junction between two semiconductors having different band gaps. A way of ensuring that the regularity of the crystal structure is continued in passing from one material to the other is to grow one substance on top of another crystal. If the two crystals have similar spacing between their atoms, then an 'epitaxial layer' can be grown, and this is the usual method of making a heterojunction.

One currently important example, used in lasers, is a junction between gallium arsenide and gallium arsenide phosphide, $GaAs\text{-}GaAs_xP_{1-x}$. Inthe latter material the As and P atoms play similar parts, and may be present in any ratio - thus varying x. A continuous range of semiconductors can thus be synthesized having steadily changing properties as x goes from 0 to 1.

An important property is the band gap, whose variation with x is shown in Fig.4.11.

The graph is made of two straight lines. As the composition is varied, every feature in the band structure varies, but at its own rate. The transition from one straight-line segment to another occurs when one local minimum in the conduction band becomes lower than another minimum. A way of describing this is to say that the transition marks the limit between GaAs-like band gap and a GaP-like band gap. As only GaAs has a direct band gap,

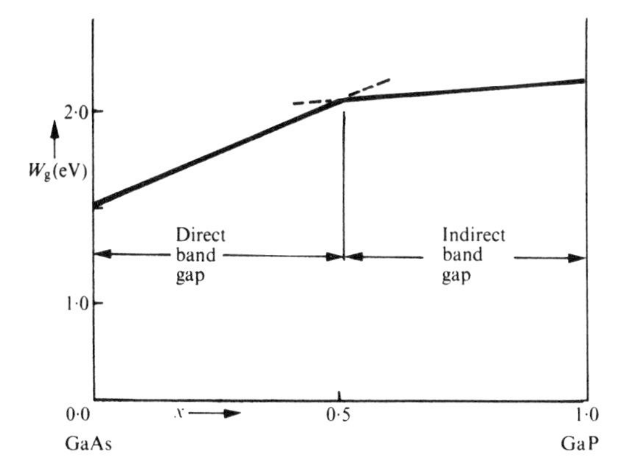

Fig.4.11. The variation of the band gap W_g of $GaAs_xP_{1-x}$ alloys with the fraction x of As.

only samples with $x < 0{\cdot}5$ are useful for the generation of light.

If there were no noticeable charge bound at the interface, the energy-band picture for heterojunctions could be constructed using the recipe for metal-semiconductor junctions, modified to allow for a step to occur in both valence and conduction bands. (See Fig.4.12).

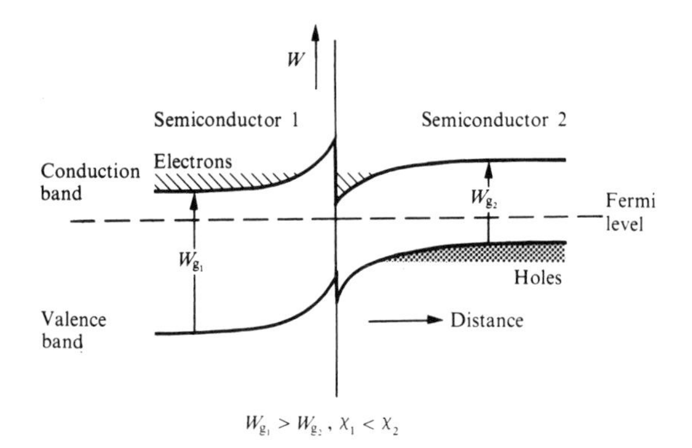

Fig.4.12. Heterojunction energy-band diagram showing discontinuity in both conduction and valence band edges.

Often one finds that all the data needed to analyse a junction is not available, and the misfit between the two lattices strains the surface, resulting in surface charges, so that the real heterojunction is difficult to analyse in simple terms.

Some heterojunction lasers and light-emitting diodes are made with adjacent layers having different band gaps. Carriers are injected into a region having a small band gap, and photons of an energy equal (more or less) to the band gap are emitted. If the photon travels in the semiconductor where it was emitted it is soon reabsorbed, but if it can reach the larger band-gap layer, it can travel with a much smaller chance of absorption, and have a greater probability of reaching the outside world.

A speculative use of heterojunctions is in light-sensitive surfaces for photomultipliers. The bending of the bands is used directly here to reduce the barrier that an electron in the conduction band must overcome to be emitted. It is possible to produce surfaces having a negative electron affinity for electrons in the conduction band of the bulk semiconductor. An electron at the bottom of the conduction band then has more energy than it needs to escape from the crystal, so if it happens to diffuse to the surface before it falls back to the valence band it will surely escape. This introduces the possibility of a surface with a high photoelectric efficiency.

PROBLEMS

4.1. The surface-charge density in an inversion layer is $1\cdot0\times10^{-5}\,\mathrm{Cm^{-2}}$. If the surface mobility of this charge is $0\cdot02\,\mathrm{m^2V^{-1}s^{-1}}$, what is the surface conductance due to this charge?

4.2. In a Si p-channel MOSFET, the gate potential is -20 V relative to the substrate whose doping density is $10^{21}\,\mathrm{m^{-3}}$, and a point A on the channel is at - 7 V. If the insulator is 10^{-6} m thick and has a relative permittivity of 6·0, what is the inversion charge at A?

4.3. In a measurement of work-function difference using the Kelvin method, (p 115), the plates of the capacitor are of an area $1\cdot0\times10^{-4}\,\mathrm{m^2}$ and are separated by an air gap of $1\cdot0\times10^{-3}$m. If the work-function difference is 2·0 eV, what is the current that flows at zero bias when one plate is vibrated at a frequency of 2000 Hz with an amplitude of 0·2 mm.

5. Systems

In this chapter there are topics which seem to be better treated in a broader environment than that of a single device, but which depend on the physics of the devices which make up electronic systems, and which profit from a physical approach to the problems.

RELIABILITY

The reliability of an electronic system is a measure of the extent to which it performs its intended function. Many devices, circuits, or systems are reliable for a period, and then fail, when they are said to have come to the end of their 'life'. A user will wish to know the expected life of a component he is purchasing, so manufacturers must ascertain the expected life, and know what must be done to produce devices with long enough lives. A related problem is that of ensuring that a device performs its intended function, i.e. 'meets its specification', when first used.

In practice, most semiconductor devices never fail; they are discarded first, so their actual life is not known. However, the fraction of the devices failing in some long interval can be quoted, and values in the range 0·01 - 0·0001 per cent per 1000 hrs might be regarded as typical. To establish such figures directly takes many devices and much time, though it is, on occasion, undertaken. By testing devices more strictly in a more severe environment, failure can be hastened, and some evidence garnered rapidly for the reliability of the device. However, relating the evidence from such an accelerated life-test to the equivalent figures under normal conditions is a difficult problem. An experimental solution is to vary the severity of the extra stress - commonly temperature - and extrapolate back to standard conditions. If, as often happens, failure occurs by some route in which an energy barrier ΔW (preferably identified) has to be overcome, then the rate should vary as $\exp(-\Delta W/KT)$. If it does so, then there is some reason for believing the extrapolation to normal conditions.

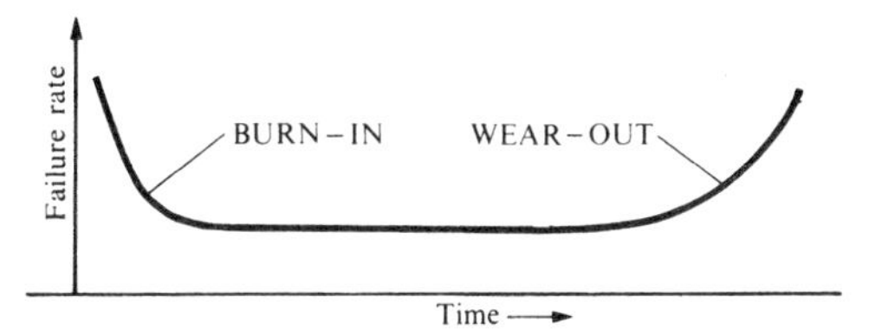

Fig.5.1. The 'bathtub curve'. For many systems, the failure rate is observed to fall to a steady value after an initial period. Later the failure rate rises as the system becomes 'worn out'.

For many devices and systems the bathtub curve (Fig.5.1) summarizes the failure rate as a function of time. There is a relatively high 'infant mortality', a lower steady failure rate, and possibly an increase in the failure rate as items become 'worn out'. The early failures can sometimes be weeded out by a burn-in test, where all devices are run at full power for a short time and then tested.

In general it may be said that perfect reliability is unattainable, but that any desired standard may be achieved by taking pains enough. When a device is newly introduced and ill understood, its manufacturing yield is low, cost is high, and reliability low. As the manufacturing techniques come under better control, the physics of the device becomes better known, the yield increases, and the balance between cost and reliability is brought even lower, though the best devices still cost most.

FAILURE MODES

The reason for the failure of a semiconductor device is unlikely to be found in the bulk semiconductor itself. Changes in the properties of the semiconductor can occur, but some such cause as bombardment by high-energy radiation is required. (The radiation can reduce minority carrier lifetime by introducing defects in the lattice which act as traps.) The p-n junction which is the heart of many semiconductor devices is also very stable, as the distinction between p- and n-type semiconductors is, speaking chemically, very slight. In evidence of this, the temperature at which dopants diffuse is much higher than the usable temperature of devices. (An exception is the very small impurity Li^+, which will stay in place in Ge and Si only while refrigerated.)

Other parts of a system offer more chances of failure or departure from specification. Interfaces between materials are particularly likely to be sites of change: e.g. semiconductor/air, semiconductor/metal, SiO_2/Si, metal/metal, metal/plastic. We consider further some examples of the many ways devices can fail, but can do no more than mention human and mechanical error in manufacture, testing, and use, which together form a major fraction of the reasons for failure.

Semiconductor/air interface

Air is reactive and variable, particularly on account of the water-vapour content, so that semiconductor surfaces cannot be exposed to it if stability is to be realized. Si forms a tough, stable, adherent oxide which goes a long way to protecting the semiconductor below it. In contrast, the compounds formed by Ge with moist air are unstable, and do not form a protective film. A contaminated surface may have a high recombination velocity and a high surface conductance, so air must be sealed away from Ge surfaces. As a result, Si has displaced Ge as the most widely used semiconductor and integrated circuits are made from Si not Ge.

Semiconductor/metal interfaces

The intended contact to a semiconductor will be an interface between it and a metal, though there may be a need to break down a thin layer of oxide in the process of making the contact. Failure can be either mechanical, the metal becoming detached from the semiconductor, or electrical, when a high-resistance region forms between metal and semiconductor. Al reduces SiO_2, and has a high electrical conductivity, so is often used to make contact to Si, being deposited by evaporation in vacuum. The surface of an integrated circuit has abrupt changes of level as a result of the sequence of manufacturing operations, and one of the difficulties in applying contacts and conductor lines is the deposition of the metal as a continuous film over the steps on the surface. Another limiting process in thin conductor strips is electromigration, where the momentum of the charge carriers in a high current density can sweep atoms of the conductor along, leading to eventual disruption of the circuit.

SiO_2/Si interfaces

The temperature coefficient of expansion of SiO_2 is much less than that of Si, so on cooling after a high-temperature oxidation, the oxide layer is strained. When the oxide layer is thick, the strain may be so severe that the oxide splits from the surface and spalls off. The effect of this depends on whether it occurs during or after manufacture. A different mode of failure occurs when there are sodium ions in or on the surface of the SiO_2. Sodium is able to move readily through SiO_2, and in a situation such as the gate insulator of a MOS transistor, there are fields - either built in or externally applied - which can drive the ions through the oxide to collect at the SiO_2/Si interface. The charge on the ions changes the gate-source bias voltage required to invert the semiconductor and create a conducting channel, so the operating point of MOS circuits can drift continuously and seriously either in use or during passive storage. Cures have included the exclusion of all Na from manufacturing equipment (human beings are good sources of Na!), and the use of other insulators (e.g. Si_3N_4) which are less permeable to Na ions.

Metal/metal interfaces

Metallurgical or chemical reactions between metals may result in changes in physical properties which affect device performance. Consequently, a physicist concerned with the failure of semiconductor devices must extend his interest to such topics. A brittle purple alloy formed between Au and Al at moderate temperatures surprised device technologists and led to many failures. Once the cause has been realized, ways of making reliable connection could be devised, for instance by the use of an intermediate layer of Mo or Ti.

Metal/plastic interfaces

The plastic resins used to encapsulate devices can cause trouble Differential expansion can fatigue and fracture the fine wires used to make contact with the integrated-circuit chip where they pass into the plastic. Lack of adhesion between leads and plastic can admit moisture, and there have been cases where plastic has decomposed into acids which have attacked the metal that was to be protected.

TECHNIQUES FOR ACHIEVING HIGH RELIABILITY

Some of the points mentioned below apply with special force to semiconductor manufacture, but many are relevant to the production of any reliable system.

The materials from which a device is made must be of very high purity - semiconductor electronics has resulted in the development of purer materials than were required for any other purpose. In addition the containers and tools used to handle the devices must be free of the most damaging contaminants. It has been found necessary to check on the checks which suppliers make to ensure the quality of their products - merely specifying the required properties is not enough. From the reciprocal point of view, it is necessary to ensure that users have the information they need to use devices effectively and safely, and that reports of failure in use are obtained and acted on.

To infer from a failed device the cause of its failure is often a piece of skilled scientific detective work, relying on a knowledge of physics, chemistry, metallurgy, engineering, and perhaps biology and psychology. This knowledge may be embodied in one man, but is more likely to represent the combined expertise of a team.

Many of the causes of failure proceed more rapidly at high temperature. A designer using electronic components will therefore need to ensure that all the components stay cool. An example which shows how much trouble is worth taking on occasion is the use of diamond as a heat-transfer element for a high-power GaAs Gunn oscillator, because certain selected diamonds have the highest known thermal conductivity at 300 K. An alternative emphasis can be given to this point by remarking that reliability may be enhanced by derating devices so they are subject to a reduced stress.

Another general method for maintaining high system reliability is by 'redundancy', so that, even though a component fails, the system keeps on working. This topic has been widely explored in both theory and practice.

Over a long manufacturing run the properties of a nominally standard device can vary considerably. However, the properties of two devices made from the same slice of Si, or even better the properties of two adjacent devices on the same chip are likely to be very similar. Where possible, then, devices should be balanced against each other and the performance of circuits made to depend on the relative rather than absolute values of components.

INTEGRATED CIRCUITS

An integrated circuit (or IC) is a chip of Si about the size of two or three letters on this page, in and on which regions of p- and n-semiconductor, insulator, and metallic conductor have been delineated photographically to make perhaps 10 or 10 000 interconnected electronic components. The way the process is carried out is discussed briefly in the next section; here we discuss some general points concerning integrated circuits.

Integrated circuits are expensive and difficult to design, but cheap to make, and as a result, many more people will be applying ICs than designing them. Some ICs will be designed for a specific purpose - e.g. as part of one manufacturer's television set - but most will perform some frequently required electronic function. An example of a linear IC is an operational amplifier, which will amplify a signal by a large factor if required, while an example of a digital IC is a random access memory. Both these are examples of MSI - medium-scale integration - where only a modest fraction of a complex system is on one chip. LSI - large-scale integration - is the use of a single chip to carry out nearly the whole of the functions of some instrument. Examples are pocket calculators and electronic clocks, where only power supplies, input, and output cannot be included on a single IC. A great advantage of LSI is that the need to 'wire up' a circuit is almost eliminated, because the transfer of signals from one part of the system to another is within the chip rather than between components.

There are a number of rival ways of making ICs, and more are being introduced. Rather than catalogue and evaluate each, some of the criteria by which an integrated circuit might be assessed are now considered.

The complexity of the particular manufacturing process limits the number of components that can be produced on one chip without there being a fault somewhere on the chip, so that for circuits with many components a simple process will be preferred, even though the potential performance is reduced. The speed of an amplifying or switching circuit may be critical - bipolar transistors are faster than MOS transistors. The power consumption should be low for battery-powered devices, and here both power when working at full speed and on standby may be important. The compatibility of a part of a complex system with the rest must be considered: does the IC in question require special power supplies; can it accept and transmit suitable signals? A process with a simple design procedure may allow users to design their own ICs, but it is possible that the increasing use of integrated circuits will render the physical understanding of electronic devices a subject for specialists only.

INTEGRATED-CIRCUIT CONSTRUCTION - THE SILICON PLANAR PROCESS

Integrated circuits are made from slices of crystallographically oriented single crystal silicon using a combination of techniques which has become known as the silicon planar process. Single components are made in the same way, a single high-power transistor occupying a larger area of Si than several ICs combined. The name 'integrated circuit' is in one sense false, as in nearly all cases the parts of the IC are familiar discrete components, separated from each other as effectively as the designer can arrange. Although there are some techniques which are peculiar to Si integrated circuits, the designer's need has been for a system whose properties could be predicted. Circuit analysis applied to the equivalent circuit of distinct components has filled this need effectively. The imperfect isolation of the intended components is described by parasitic equivalent circuit elements, unwanted but unavoidable.

There are many ways of making integrated circuits which combine the same processes in varying orders. Here we present one simplified version.

The starting point of our process is a slice of p-type Si, often 50 mm in diameter (Fig.5.2(a)), which is polished and etched to produce a flat undamaged surface. A large number of identical devices are made on one slice, in a way to be described.

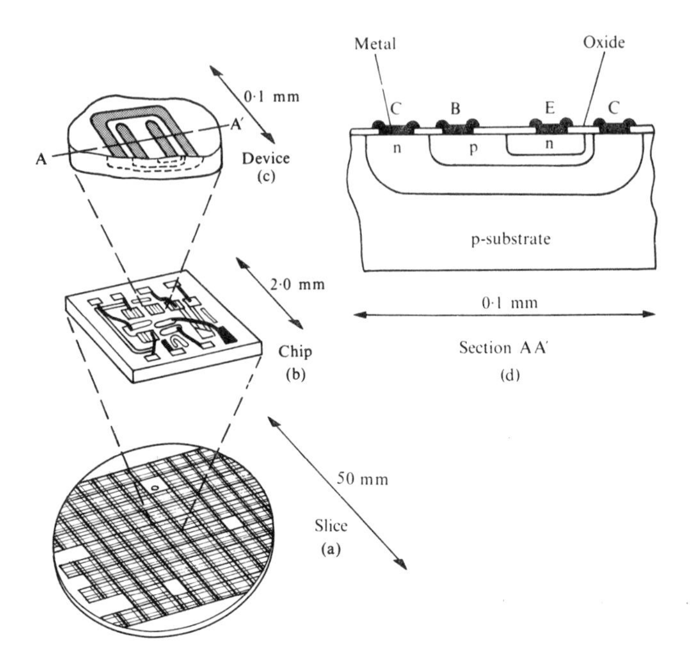

Fig.5.2. Integrated circuits: (a) a slice of single-crystal Si; (b) one IC chip; (c) the conductor pattern on top of one transistor forming part of the IC in (b); (d) a cross-section through the n-p-n transistor on the p-type substrate.

After testing, the slice is scribed in two directions with a diamond point, so that the chips (Fig.5.2(b)) can be separated. The chips are alloyed onto larger metal bases, and fine Au wires pressure-bonded to make connections from aluminium contacts on the chip to leads on the base. The IC is finally encapsulated in plastic or some other hermetic package.

The individual IC patterns are defined photographically as sketched in Fig.5.3(a). The clean slice is prepared by first growing a layer of SiO_2 on its surfaces by heating in an atmosphere containing O_2 or H_2O, and then covering with a thin layer of photoresist, a photosensitive material which hardens when exposed to ultraviolet light. A mask, which contains many

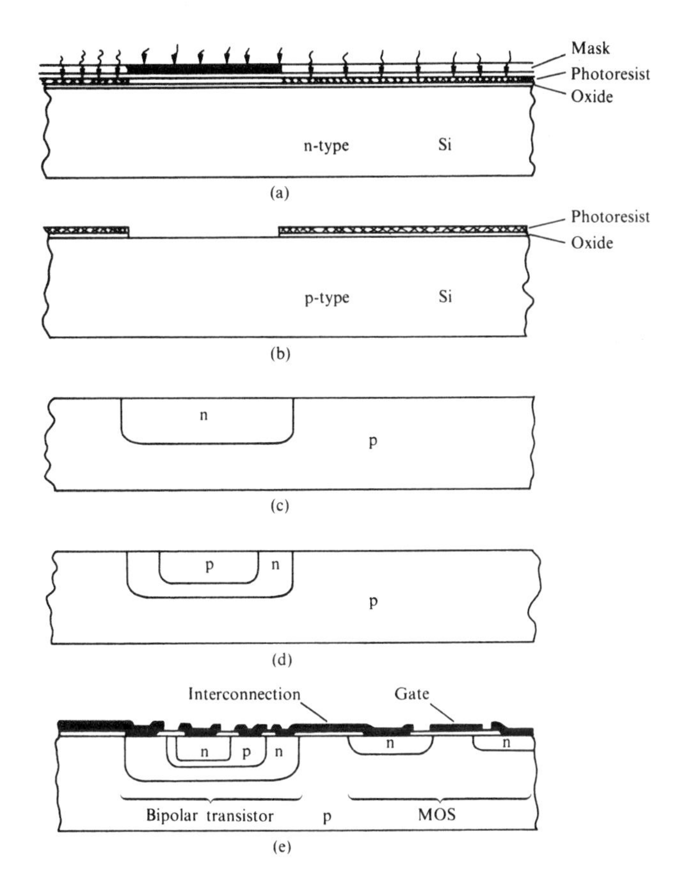

Fig.5.3. Stages in making an IC: (a) defining patterns by illuminating photoresist by ultraviolet light through a mask; (b) photoresist and underlying SiO_2 etched away; (c) an n-type region diffused into the p-type substrate; (d) a further n-type region diffused into the p-type substrate; (e) the completed process, after a further diffusion, oxidation, and metallization. On the right are two n-type regions forming an n-channel MOS.

copies of the negative of the required pattern, is meanwhile prepared by photographic reduction of a large master pattern. The mask is placed in contact with the slice of Si, the exposed areas of photoresist hardened by a flash of ultraviolet light, and the rest dissolved in a solvent. The areas of SiO_2 thus exposed are dissolved in HF, to which both the photoresist and Si are resistant (Fig.5.3(b)). The remaining photoresist is removed and the slice is now ready for diffusion of P donors (see pp 14-15) at high temperature, resulting in patterns of n-type semiconductor embedded in the p-type substrate, as sketched in Fig.5.3(c). The sequence of oxidation, covering with photoresist, exposure through a mask, etching, and diffusion is repeated twice more to introduce further p- and n-type layers for the base and emitter regions in Fig.5.3(d) and 5.3(e). In Fig.5.3(e) n-regions for the source and drain of an n-channel MOS have also been deposited. This makes the point that MOS transistors are markedly simpler to make than are bipolar transistors.

A layer of oxide to protect the surface and act as the insulator for the MOS transistor is grown and holes etched through it to allow contact to be made to the circuit. Finally, Al is evaporated and etched to the pattern of interconnections and contact pads as in Fig.5.3(e).

Fig.5.2(c) and (d) show the relation between the surface patterns and the distribution in depth of the emitter, base, and collector regions for a bipolar transistor, which itself might be repeated many times on a single IC. The shape of the collector contact allows current to reach the contact from the active regions of the collector by two parallel paths, thus reducing the resistance.

The edge of a pattern can be positioned to an accuracy of about 1 μm, but when allowance has been made for the need to align up to half a dozen successive patterns, the minimum width of a component has to be much more. A resistor could be 10 μm in width, while a bipolar transistor may occupy an area 100 μm 70 μm. The depths of the components are determined by the depths to which impurities diffuse, which is controlled by the time and temperature of the diffusion stages of processing. A first diffusion might reach 2-3 μm, with later doses of dopant producing junctions spaced by 0·2 - 1·0 μm.

Many but not all conventional components can be produced in IC form. Junction diodes and the various forms of transistors can be readily produced. Resistors can be made conveniently so long as the value of resistance required is neither very high nor very low. Small- to medium - size capacitors can be made in two ways - either by using the depletion-layer capacity of a p-n junction or by using the surface SiO_2 as a dielectric between two plates, one formed by the semiconductor and the other by the aluminium used for conductors.

Attempts to devise effective ways of making inductors and transformers using integrated-circuit techniques have not been successful. The alternative approach, of devising circuits which can perform the required functions without the use of magnetic components, has perhaps had rather greater success. For example, several semiconductor methods have been developed to rival the ferrite toroidal core array for storing binary information.

The sequence of operations which has been described has been elaborated in several ways in practice. Epitaxial growth of Si with a low impurity density and hence a high resistivity considerably increases the flexibility of the manufacturing process, and allows higher performances to be achieved. Au is found to diffuse rapidly and far via interstitial sites, and acts as a recombination centre, so it is used when short minority lifetimes are wanted.

The main technique used to ensure isolation between components is junction isolation, when every component is separated from the substrate by a reverse-biased p-n junction. To ensure that the junction is reverse-biased, a p-type substrate must be taken to the most negative potential in the circuit. Other methods of isolation involving layers of SiO_2 or other insulator have been tried, but their increased effectiveness has not outweighed their greater difficulty of manufacture.

The relative cost of various kinds of components has turned out to be different in ICs from the ratios which apply to discrete components. The design of IC circuits reflects this change. Diodes and transistors are used freely, and resistors and capacitors avoided where possible. An example of this process taken to its logical conclusion is the COSMOS or COmplementary Symmetry MOS family of ICs, where each stage of the circuits has an n-channel and p-channel MOS in series. The devices are made so that both their gates may be joined and controlled from a single input signal (Fig.5.4). When one device is ON the other is OFF, and the distinction between the active device and its passive load is lost.

Use is made of the very similar temperature of adjacent devices on the same chip to balance the temperature dependence of the properties of one component against those of a similar component.

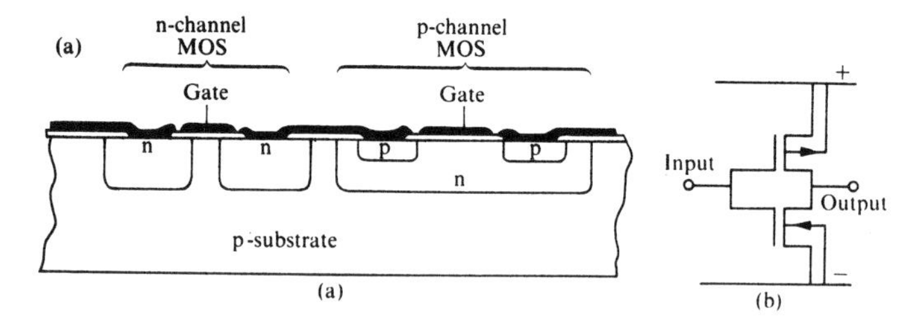

Fig.5.4. COSMOS: (a) schematic construction; (b) an n-channel MOS and a p-channel MOS connected to act a logical inverter; when one is ON, the other is OFF.

In order to avoid coupling signals from one stage of an amplifier to another via the power supply, a discrete-component circuit would use a large capacitor to smooth the flow of current. The integrated-circuit solution uses a complementary circuit whose current falls as the current in the main circuit rises, the extra transistors required being easier to include than a large capacitor.

NOISE

Noise in colloquial terms is any unwanted signal. Some noise comes from sources outside the circuit (lightning, solar flares, or petrol ignition sparks) and with this we shall not be further concerned. In digital circuits, signals applied on one lead may induce signals on other leads, and a circuit designer has to ensure that the circuit can distinguish reliably between such cross-talk and intended signals. Our attention however, will be concentrated on the irreducible minimum of noise in all circuits which is dependent on first the non-zero temperature of the system, and second the finite magnitude of the charge on an electron.

If a fluctuating current i has a mean value $\bar{i}$, then the fluctuation in i is Δi and $\Delta i = i - \bar{i}$. Now $\overline{\Delta i} = 0$, so $\overline{\Delta i}$ is useless as a measure of the noisiness of i, but $\overline{(\Delta i^2)} \neq 0$, and is known as the variance of i, written var(i). Then

$$\text{var}(i) = \overline{(i - \bar{i})^2} = \overline{i^2} - (\bar{i})^2$$

(Check this by expanding the middle term).

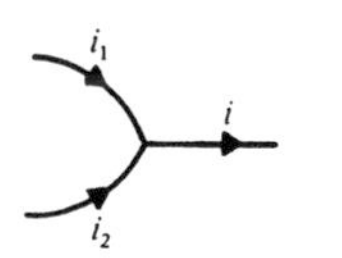

If now a current i has two components i_1 and i_2, which perhaps have been flowing in separate wires and now join, the variance of i is

$$\text{var}(i) = (\Delta i^2) = (\Delta i_1 + \Delta i_2)^2 = \Delta i_1^2 + \Delta i_2^2 + 2\,\Delta i_1 \Delta i_2 \ .$$

If Δi_1 and Δi_2 are 'uncorrelated', then the third term is zero, so that variances add directly. If the sources of noise are causally independent, then the noise is not correlated. If there is some correlation then $\Delta i_1 \Delta i_2$ may be positive or negative. (Exercise: find var(i) when $\Delta i_1 = \Delta i_2$ and when $\Delta i_1 = -\Delta i_2$).

A particular kind of randomness, and hence noise, comes from a train of similar independent events. If the average number of events in a sample is $\bar{n}$, then a result of statistical theory is that noise may be spread over a wide range of frequency, and we use the concept of spectral intensity of the fluctuations in some quantity n in a frequency range $f + \delta f$, which is written $S_n(f)$, where $S_n(f)\delta f$ is the average value of the square of the fluctuations in n between f and $f + \delta f$. A further result of statistical theory is that $S_n(f) = 2\,\mathrm{var}(n)$, so that for our random series $S_n(f) = 2\,\bar{n}$.

A current I of I/e independent elctrons per second would have a variance I/e, so the spectral intensity of the fluctuations in the number of electrons per second would be $S_n(f) = 2I/e$, while the spectral intensity of the fluctuations in the current would be

$$S_I(f) = e^2 S_n(f) = 2eI. \tag{5.1}$$

The name shot noise is given to this kind of electrical noise.

The idea of thermal noise applies properly to systems in thermodynamic equilibrium, but may be applied safely to many systems so long as they are not perturbed strongly away from equilibrium. A resistor through which the mean current is zero is a good example of a situation which can be taken to be in thermal equilibrium.

An argument based on classical thermodynamics leads to a set of expressions for the spectral intensity of fluctuations in the current, voltage, or power in a resistance R at temperature T

$$S_I(f) = 4\,\kappa T/R,$$
$$S_V(f) = 4\,\kappa TR, \qquad (5.2)$$
$$S_P(f) = 4\,\kappa T\,.$$

The quantized version of (5.2), including the effect of zero-point energy is

$$S_V(f) = 4\,hfR\{\tfrac{1}{2} + 1/(\exp(hf/\kappa T)-1)\} \qquad (5.3)$$

(Check that (5.3) reduces to (5.2) at low frequency, i.e. when $hf \ll \kappa T$).

The simple versions (5.2) are generally used. They show a flat spectrum for all f. From a practical point of view (5.2) sets a limit below which the noise from a resistor cannot fall, though poor resistors may well have a higher noise output.

Consider now the low-frequency noise properties of a p^+-n junction diode which obeys the equation $I = I_R\{\exp(eV/\kappa T)-1\}$. When the diode current is I, the hole current injected into the n-side will be $I + I_R$, while the hole current leaking back into the p-side will be $-I_R$. In both of these currents the holes move independently, and contribute full shot noise; for the injected current because the process is random diffusion, and for the leakage because the arrival of holes at the edge of the depletion layer is random, and nothing can arrest their motion once in the high field of the depletion layer. Electron currents are being ignored because the diode is p^+-n not p-n.

The full spectral intensity of current fluctuation on the basis of each current contributing according to equation (5.1) is

$$S_I(f) = 2\,eI + 4eI_R. \qquad (5.4)$$

We can compare this with the result derived from the slope resistance of the diode when $I \gg I_R$. The resistance is $\kappa T/eI$, so that from eqn (5.2) $S_I(f)$ is predicted to be $4eI$. The distinction is important, the slope resistance of a p-n junction diode does not act as a thermodynamic noise generator; it generates only half the noise. Extra effects have to be taken into account at high frequency, but for simplicity we analyse only low-frequency effects, both for the junction diode and for the bipolar transistor.

For our model of a p-n-p transistor we start with the Ebers-Moll equation (3.31). Using the relation $\alpha_i I_{CBo} = \alpha_n I_{EBo}$, the equations may be rewritten for the normal bias where V_{EB} is small and positive and V_{CB} is large and negative,

$$I_E = \{\alpha_n I_{EBo} \exp(eV_{EB}/\kappa T)\} + \{(1 - \alpha_n I_{EBo})\exp(eV_{EB}/\kappa T)\} - \{I_{EBo}(1 - \alpha_n)\},$$

$$I_C = \{-\alpha_n I_{EBo} \exp(eV_{EB}/\kappa T)\} - \{I_{CBo}(1 - \alpha_i)\}, \qquad (5.5)$$

$$I_B = -(I_E + I_C).$$

There are four distinct terms in braces in eqn (5.5), and they have been indicated by the four groups of holes flowing in Fig.5.5(a). Each carries full shot noise, and is uncorrelated at low frequencies with the others. The equivalent circuit, Fig.5.5(b), includes the noise current generators from Fig.5.5(a), shown in dotted circles, and ordinary circuit elements (the low-frequency hybrid-π model), which operate on signals or noise alike.

The values of i_{n3} and i_{n4} are small in Si transistors. When they may be ignored, the value of the other noise currents is

$$i_{n1}^2 = S_{i1}(f)df = 2e\,\alpha_n I_{EBo} \exp(eV_{EB}/\kappa T)\delta f = 2eI_C\delta f,$$

$$i_{n2}^2 = S_{i2}(f)df = 2e(1 - \alpha_n)I_{EBo} \exp(eV_{EB}/\kappa T)df = 2eI_B df.$$

From the derivation of the hybrid-π parameters, (pp 72-75) $r_\pi = \kappa T/eI_B$, so that, in agreement with our analysis of a junction diode

$$i_{n2}^2 = (2\kappa T/r_\pi)\delta f.$$

A simple circuit suitable for analysis, including a source resistor with its own noise generator, is shown in Fig.5.5(c).

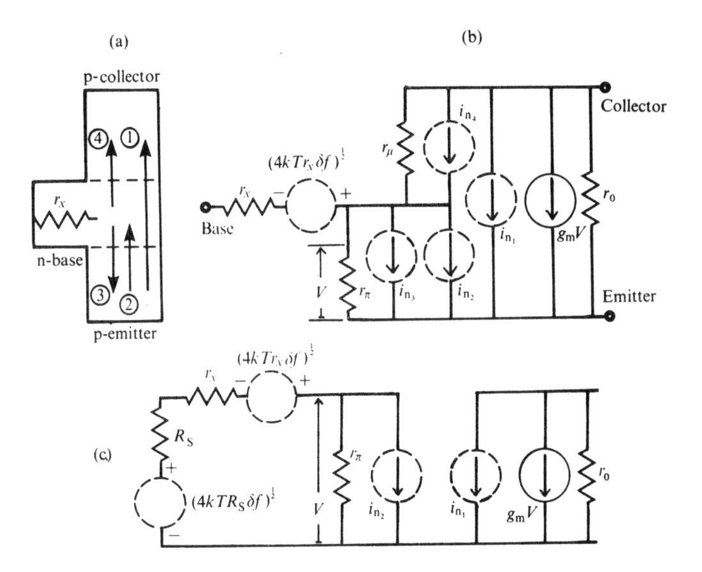

Fig.5.5. Bipolar transistor noise: (a) hole currents in a p-n-p transistor corresponding to eqn(5.5): (b) the equivalent circuit of (a): (c) a simplified equivalent circuit of the transistor and a signal source.

Before we can carry out the analysis, we must consider how the noise in a circuit should be assessed.

NOISE MEASURES

There have been a number of parameters used to describe the noise in a circuit; we shall study the noise figure; other names that have been used are 'operating noise figure', 'operating noise factor', and 'spot noise figure'. The noise figure F for a circuit is the ratio of the total noise power delivered to the output of the circuit when the source resistance is at 290 K to the part of the noise power at the output which is due to the source resistor. The noise figure is thus always greater than one, though in a good circuit it may be only a little greater.

One way to measure the noise figure is indicated in Fig.5.6. A calibrated noise source, such as a temperature-limited vacuum

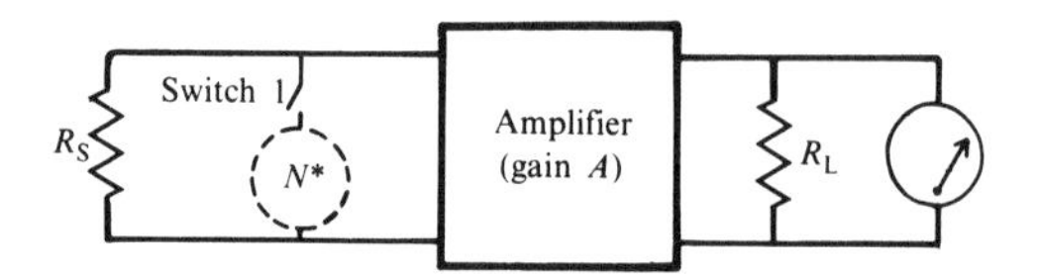

Fig.5.6. To measure the noise figure F, the output noise with switch 1 open is noted, and then switch 1 is closed and the calibrated noise source N^* adjusted until the output noise power doubles.

diode is used. First note the noise power arriving at R_L when the calibrated noise source is switched off. Second, increase the current in the noise source until the power in R_L is doubled. If the power in R_L in the first case is O_1, then

$$O_1 = AN_S + AN_A$$

where N_S is the noise power from R_S and N_A is the noise power from the amplifier, which we describe as if it were all produced at the input. When the noise power arriving at R_L is doubled, we have

$$2\,O_1 = A(N_S + N_A) + AN^* ,$$

where N^* is the noise power from the calibrated source. Then $N^* = N_A + N_S$, and from the definition of F

$$F = A(N_A + N_S)/AN_S = N^*/N_S .$$

The value of N^* is known from the calibration of the instrument, and N_S is known from the resistance of R_S and its temperature.

The noise figure for the transistor amplifier of Fig.5.5(c) can now be calculated, and from the calculation some guidance obtained for the best ways of making a low-noise amplifier.

The noise sources associated with R_S, r_x, and r_π produce a total (noise voltage)^2 of V^2 across r_π, where

$$V^2 = \frac{4\kappa T\ \delta f\ (R_S + r_x)r_\pi{}^2}{(R_S + r_x + r_\pi)^2} + \frac{2\kappa T\ \delta f}{r_\pi}\ \frac{(R_S + r_x)^2 r_\pi{}^2}{(R_S + r_x + r_\pi)^2} .$$

Across the output resistance r_O, the square of the noise voltage due to all sources is V_{out}^2, where

$$V_{out}^2 = r_O^2(2eI_C\ \delta f + g_m^2 V^2). \tag{5.6}$$

When only the noise from R_S is considered

$$V_{out}^2 = r_O^2 g_m^2 \frac{4KT\ \delta f\ R_S r_\pi^2}{(R_S + r_x + r_\pi)^2} \tag{5.7}$$

The value of F is the ratio of eqns (5.6) to (5.7), and, in writing it out, $g_m r_\pi$ has been replaced by β, the small signal current gain.

$$F = 1 + \frac{r_x}{R_S} + \frac{(R_S + r_x)^2}{2R_S r_\pi} + \frac{e\ I_C(R_S + r_x + r_\pi)^2}{2\beta^2 KT\ R_S} \tag{5.8}$$

This result would be the starting point for optimizing the noise properties of a low frequency common-emitter amplifier. The best values of R_S and I_C for a given transistor and the best value of β can be predicted. It would of course be worthwhile comparing experimental results with theory to show whether the terms which have been ignored in the approximation can be validly omitted.

When passing current all devices have been found to have low-frequency noise power in excess of the value predicted by $4\ KT\ df$. This is called flicker, or $1/f$ noise. The corner frequency at which the graph of the spectrum bends up from a f^0 dependence to an f^{-1} dependence may be in the range of 100 Hz to 10 000 Hz, and the f^{-1} law appears to hold to the lowest frequencies tested (10^{-5} Hz). The random filling and emptying of a distribution of states at semiconductor surfaces is thought to be the usual cause of this noise, and, unfortunately, as manufacturing process control improves for a device, the corner frequency for $1/f$ noise moves steadily down. All that can be done by a user to cope with $1/f$ noise where a small, low-frequency signal is to be amplified is to select low-noise, high-quality devices.

PROBLEMS

5.1. Design an integrated circuit resistor to have a resistance of 150 Ω. The resistor is to be isolated using junction isolation, the standard doping density for both p- and n-regions is to be 10^{22} m^{-3}, and the minimum depth or width of a patterns is 5×10^{-6} m.

5.2. Check the parasitic depletion-layer capacitance of your resistor (not forgetting that it has sides), assuming that the junction has 2·5 V reverse bias, and hence find its high-frequency limit (from CR).

5.3. Find $(S_V(f)\delta f)^{\frac{1}{2}}$ and $(S_I(f)\delta f)^{\frac{1}{2}}$ for a 1 MΩ resistor and a 1 Ω resistor, taking δf to be 1 MHz and the temperature to be 300 K.

5.4. Plot the way F varies in eqn (5.8) as R_S is varied, and hence find an optimum value of R_S. Take r_x = 100 Ω, r_π = 1250 Ω, I_C = 1 mA, β = 50, T = 300 K.

5.5. Take a device and list the interfaces between materials where instability could lead to malfunction of the device.

Appendix: Selected properties of semi-conductors at 300K.

The following properties of semiconductors are tabulated:-

- W_g - Band gap in eV
- I or D - Direct or indirect gap
- n_i - Intrinsic carrier concentration at 300 K in m^{-3}
- μ_e - Electron mobility in $m^2V^{-1}s^{-1}$
- μ_h - Hole mobility in $m^2V^{-1}s^{-1}$
- $\frac{m_e}{m_o}$ - Ratio of electron effective mass to free electron mass
- $\frac{m_h}{m_o}$ - Ratio of hole effective mass to free electron mass
- ϵ_r - Relative permittivity
- T_m - Melting point in K

	Si	Ge	GaAs	InSb
W_g	1.1	0·7	1·4	0·17
I or D	I	I	D	D
n_i	$1{\cdot}6 \times 10^{16}$	$2{\cdot}5 \times 10^{19}$	$1{\cdot}2 \times 10^{13}$	$1{\cdot}3 \times 10^{22}$
μ_e	0·14	0·39	0·65	7·8
μ_h	0·05	0·19	0·04	0·4
$\frac{m_e}{m_o}$	1·0	0·22	0·07	0·13
$\frac{m_h}{m_o}$	0·59	0·37	0·51	0·18
ϵ_r	12	16	12	18
T_m	1693	1210	1553	796

Answers to selected problems

1.3 0·4, 8×10^{-11}, $1{\cdot}3 \times 10^{-8}$.

1.4 For electrons, $5{\cdot}3 \times 10^{-8}$ m or 230 interatomic spaces.
For holes, $1{\cdot}4 \times 10^{-8}$ m or 60 interatomic spaces.

2.1 18·6 Ω, 0·054 A, $1{\cdot}2 \times 10^{10}$ W m^{-3}.

2.3 5×10^{-7} s.

2.4 $4{\cdot}6 \times 10^{21}$ m^{-3}, 2·6 T.

3.1 0·51 V, $1{\cdot}2 \times 10^{-4}$ m in the n-region, $1{\cdot}2 \times 10^{-5}$ in the p-region, 0·008 pF, 3000 V.

3.2 Ratio is 1·6.

3.3 32 μs.

3.5 For the Si transistor $\beta = 120$, $r_0 = 2{\cdot}5$ k Ω,
$g_\pi = 8$ mA V^{-1}, $g_m = 1{\cdot}0$ A V^{-1}.
For the Ge transistor $\beta = 35$, $r_0 = 1{\cdot}3$ k Ω,
$g_\pi = 15$ mA V^{-1}, $g_m = 1{\cdot}0$ A V^{-1}: The values for g_π and g_m assume that simple theory applies validly.

3.7 0·0027 V.

4.1 2×10^{-7} S per square.

4.2 $1{\cdot}4 \times 10^{-4}$ C m^{-2}.

4.3 4×10^{-9} A.

5.3 For 1 Ω, 4×10^{-8}A, 4×10^{-8}V; for 1 MΩ, 4×10^{-11}A, 4×10^{-5}V.

5.4 Minimum F is 1·54 at $R_S = 540$ Ω.

Bibliography

There follows a small selection (mostly recent) containing many further references.

BEALE, J.R.A., EMMS, E.T., and HILBOURNE, R.A., (1971) *Microelectronics*. Taylor and Francis, London. Describes the Si planar process and illustrates the uses of devices.

CARROLL, J.E. (1974) *Physical models for semiconductors*. Arnold, London. An undergraduate text describing device physics.

JOWETT, C.E. (1966) *Reliability of electronic components*. Illiffe Books, London. A practical summary.

KANO, K. (1972) *Physical and solid state electronics*. Addison-Wesley, New York. A good teaching text, as the points are made thrice. Some diagrams are suspect.

PANCOVE, J.I. (1971) *Optical processes in semiconductors*. Prentice-Hall, New York. A treatment at postgraduate level.

RICHMAN, P. (1973) *MOS field-effect transistors and integrated circuits*, Wiley-Interscience, New York. Good on MOS device physics.

ROSENBERG, H.M. (1974) *Solid state physics (OPS 9)*. Clarendon Press, Oxford. Deals more thoroughly with the material of chapter 1.

SEEC Series. (Semiconductor Electronics Education Committee) *Vols.1-7*. (1964) John Wiley, New York. A valuable series dealing with junction transistors in detail. The section on noise in the present text follows the treatment in Vol.4.

SMITH, R.A. (1959) *Semiconductors*. Cambridge University Press. Often used as a basic source by the authors of more recent works.

SZE, S.M. (1969) *Physics of semiconductor devices*. Wiley-Interscience, New York. An advanced test with much technical information.

VAN DER ZEIL, A. (1968) *Solid state physical electronics*. Prentice-Hall, New York. An authoritative handbook, densely packed, at an advanced undergraduate level.

Index

Physical constants and conversion factors

Avogadro constant	L or N_A	$6{\cdot}022 \times 10^{23}\ \text{mol}^{-1}$
Bohr magneton	μ_B	$9{\cdot}274 \times 10^{-24}\ \text{J T}^{-1}$
Bohr radius	a_0	$5{\cdot}292 \times 10^{-11}\ \text{m}$
Boltzmann constant	k	$1{\cdot}381 \times 10^{-23}\ \text{J K}^{-1}$
charge of an electron	e	$-1{\cdot}602 \times 10^{-19}\ \text{C}$
Compton wavelength of electron	$\lambda_C = h/m_e c$	$= 2{\cdot}426 \times 10^{-12}\ \text{m}$
Faraday constant	F	$9{\cdot}649 \times 10^{4}\ \text{C mol}^{-1}$
fine structure constant	$\alpha = \mu_0 e^2 c/2h$	$= 7{\cdot}297 \times 10^{-3}$ ($\alpha^{-1} = 137{\cdot}0$)
gas constant	R	$8{\cdot}314\ \text{J K}^{-1}\ \text{mol}^{-1}$
gravitational constant	G	$6{\cdot}673 \times 10^{-11}\ \text{N m}^2\ \text{kg}^{-2}$
nuclear magneton	μ_N	$5{\cdot}051 \times 10^{-27}\ \text{J T}^{-1}$
permeability of a vacuum	μ_0	$4\pi \times 10^{-7}\ \text{H m}^{-1}$ exactly
permittivity of a vacuum	ϵ_0	$8{\cdot}854 \times 10^{-12}\ \text{F m}^{-1}$ ($1/4\pi\epsilon_0 = 8{\cdot}988 \times 10^{9}\ \text{m F}^{-1}$)
Planck constant	h	$6{\cdot}626 \times 10^{-34}\ \text{J s}$
(Planck constant)/2π	$\hbar$	$1{\cdot}055 \times 10^{-34}\ \text{J s} = 6{\cdot}582 \times 10^{-16}\ \text{eV s}$
rest mass of electron	m_e	$9{\cdot}110 \times 10^{-31}\ \text{kg} = 0{\cdot}511\ \text{MeV}/c^2$
rest mass of proton	m_p	$1{\cdot}673 \times 10^{-27}\ \text{kg} = 938{\cdot}3\ \text{MeV}/c^2$
Rydberg constant	$R_\infty = \mu_0^2 m_e e^4 c^3/8h^3$	$= 1{\cdot}097 \times 10^{7}\ \text{m}^{-1}$
speed of light in a vacuum	c	$2{\cdot}998 \times 10^{8}\ \text{m s}^{-1}$
Stefan–Boltzmann constant	$\sigma = 2\pi^5 k^4/15h^3c^2$	$= 5{\cdot}670 \times 10^{-8}\ \text{W m}^{-2}\ \text{K}^{-4}$
unified atomic mass unit (^{12}C)	u	$1{\cdot}661 \times 10^{-27}\ \text{kg} = 931{\cdot}5\ \text{MeV}/c^2$
wavelength of a 1 eV photon		$1{\cdot}243 \times 10^{-6}\ \text{m}$

1 Å $= 10^{-10}$ m; 1 dyne $= 10^{-5}$ N; 1 gauss (G) $= 10^{-4}$ tesla (T);
0°C $= 273{\cdot}15$ K; 1 curie (Ci) $= 3{\cdot}7 \times 10^{10}\ \text{s}^{-1}$;
1 J $= 10^{7}$ erg $= 6{\cdot}241 \times 10^{18}$ eV; 1 eV $= 1{\cdot}602 \times 10^{-19}$ J; 1 $\text{cal}_{th} = 4{\cdot}184$ J;
$\ln 10 = 2{\cdot}303$; $\ln x = 2{\cdot}303 \log x$; $e = 2{\cdot}718$; $\log e = 0{\cdot}4343$; $\pi = 3{\cdot}142$